Essentials of Plant Breeding

Essentials of
Plant Breeding

K.V. MOHANAN
Professor
(Genetics and Plant Breeding Division)
Department of Botany
University of Calicut
Kerala

PHI Learning Private Limited
New Delhi-110001
2010

Rs. 4[illegible]57-

ESSENTIALS OF PLANT BREEDING
K.V. Mohanan

ISBN-978-81-203-3968-2

Published by Asoke K. Ghosh, PHI Learning Private Limited, M-97, Connaught Circus, New Delhi-110001 and Printed by Mudrak, 30-A, Patparganj, Delhi-110091.

Contents

Preface

Plant breeding is the art and science of the genetic improvement of crops to produce new varieties with increased productivity and quality. The science of plant breeding evolved in the hands of the great farmers of the world for centuries. In the twentieth century, scientists specialized in crop genetics and improvement working around the globe refined the methodologies of this branch of science. In recent years, the techniques used have rapidly advanced to genetic and molecular levels. As a result, new varieties of crop plants, medicinal plants and ornamentals have been developed.

The present book has been written as a course material that can be adopted both at the undergraduate and postgraduate levels. It will also prove useful to professionals interested in plant science and crop science with a basic knowledge of crop genetics.

The book comprises eight chapters. The first chapter gives an overview of plant breeding. The second chapter builds a preliminary foundation by providing knowledge on the basic concepts of biology, particularly the principles of genetics and biostatistics. The third chapter deals with the conventional methods of plant breeding, whereas the fourth chapter emphasizes on modern methods. The fifth chapter discusses different breeding methods used for pest and disease resistance, quality improvement, heterosis breeding, ideotype breeding, etc. The sixth chapter has a special mention on variety release and seed certification. The seventh chapter discusses the importance of Intellectual Property Rights (IPR) in plant breeding and the last chapter adds a note on sustainable agriculture.

Special care has been taken to make the book suitable and useful to students. The author invites suggestions from readers for the improvement of text.

K.V. Mohanan

Acknowledgements

The author has been teaching Plant Genetics and Breeding at postgraduate level for the last twenty-eight years. The credit for the development of the approach followed in this book goes to the students who attended his lectures critically and raised genuine questions.

The research scholars who worked under his supervision for their postgraduate and doctoral programmes, many of whom are scientists and teachers presently, have also contributed in formulating the concept and approach that has been adopted in this book.

However, the content has been developed based on the requirements of the students. The author records his deep sense of gratitude to all earlier authors who have contributed important reference books to the subject of plant breeding.

Finally, he would like to express his most sincere thanks to PHI Learning, New Delhi, for giving him the opportunity to write this book.

K.V. Mohanan

CHAPTER 1

Plant Breeding An Overview

The evolution of human civilization started from the primitive cave dwelling hunter gatherers. Their societies lived a nomadic lifestyle in which they moved from place to place in search of food and shelter. But the knowledge that seeds can be sown to grow food crops changed their lifestyle completely. Human culture took a significant turn and the primitive human societies started living at a particular place at least up to the harvest of the crop from the seeds that were sown by them. This denotes the dawn of agriculture. The evolution of human culture and civilization from this stage till today was steered by the evolution of agricultural systems and practices. That is why it is rightly said that agriculture is the mother of culture.

1.1 MAJOR CROP RESEARCH CENTRES

The growth of agriculture promoted the evolution of techniques to improve the seed stocks and propagules of cultivated plants through the continuous efforts of the farming community. The art and science of improving cultivated plants so as to achieve better varieties is called *plant breeding*. Plant breeding has been a social activity right from the beginning of agriculture till the 19th century. In the 20th century, the development of the science of genetics enhanced the scope and objectives of plant breeding and as a result, it became an organized scientific activity. Research institutes and laboratories were established throughout the world to organize and co-ordinate agricultural research. Presently, the Consultative Group for International Agricultural Research (CGIAR), an organization established under FAO, co-ordinates agricultural research on a global basis. Under the aegis of CGIAR, the following international research institutes are functioning presently (Table 1.1).

Table 1.1 Crop research institutes under CGIAR

IRRI	International Rice Research Institute, Manila, The Philippines
CIMMYT	International Wheat and Maize Improvement Centre, Elbaton, Mexico
ICRISAT	International Crop Research Institute for Semi-Arid Tropics, Hyderabad, India
IITA	International Institute for Tropical Agriculture, Ibadan, Nigeria
CIAT	International Centre for Tropical Agriculture, Palmira, Colombia
CIP	International Potato Centre, Lima, Peru
WARDA	West African Rice Development Research Station, Monrovia, Liberia
BIODIVERSITY INTERNATIONAL	Biodiversity International, Rome, Italy (Previously known as IBPGR and IPGRI)
ICARDA	International Centre for Agricultural Research in Dry Areas, Lebanon, Syria

In India, plant breeding research is being carried out mainly by institutes, universities and research stations spread around the country. The system can be classified into institutes functioning under Indian Council of Agricultural Research (ICAR), Ministry of Agriculture, institutes functioning under commodity boards, agricultural universities and other institutions including quasi government and private research stations and conventional universities. Crop research institutes carrying out research on almost all major crops have been established by ICAR (Table 1.2).

Four commodity boards have been instituted by Government of India under the Ministry of Commerce for the four plantation crops. Crop research institutes have been established by the four commodity boards for the plantation crops (Table 1.3).

Table 1.2 Some major research institutes functioning under ICAR

IARI	Indian Agricultural Research Institute	New Delhi
NBPGR	National Bureau of Plant Genetic Resources	New Delhi
CTCRI	Central Tuber Crop Research Institute	Trivandrum
CPCRI	Central Plantation Crop Research Institute	Kasaragod
IISR	Indian Institute of Spices Research	Calicut
NRC-OP	National Research Centre for Oil Palm	Pedavegi
NRCC	National Research Centre for Cashew	Puthur
CRRI	Central Rice Research Institute of India	Cuttack
NRC for Sorghum	National Research Centre on Sorghum	Hyderabad
SBI	Sugarcane Breeding Institute	Coimbatore

Table 1.3 Research institutes functioning under commodity boards

Name of the crop	*Name of the board*	*Name of the institute*	*Location*
Rubber	Rubber Board, Kottayam	Rubber Research Institute of India	Kottayam
Cardamom	Spices Board, Cochin	Indian Cardamom Research Institute	Myladumpara, Idukki
Coffee	Coffee Board, Bangalore	Central Coffee Research Institute	Chikmagalore
Tea	Tea Board, Kolkata	Tea Research Institute	Tocklai

1.2 METHODS OF PLANT BREEDING

The major plant breeding activities include domestication of plants, introduction of plants, acclimatization, selection of plants, plant hybridization, mutation breeding, polyploidy breeding and biotechnological breeding which are carried out for different purposes. Different techniques have been developed based on the modes of reproduction of crop plants.

1.2.1 Domestication of Plants

In nature, the origin of plants took place in the wild condition. During the course of evolution of agricultural practices, man began to grow some plants under his control. This process of bringing wild plants under cultivation is called *plant domestication.*

1.2.2 Introduction of Plants

Besides the domestication of wild plants, introduction of plants from other continents, countries, geographical regions, etc. to new areas of cultivation is also an important process in plant breeding. This technique is called *plant introduction.* In other words, the process of introducing new plants from the place of its origin or cultivation to a place with different climatic conditions is called *plant introduction.*

1.2.3 Acclimatization

The physiological adaptation of plants to climatic or environmental changes, such as light, soil, temperature or altitute is known as *acclimatization.*

1.2.4 Selection of Plants

Plants, both domesticated as well as introduced, show considerable degrees of variations with respect to different characters. Some of these plants are superior whereas the others are inferior in performance. The process of selection of superior plants is an important method for the improvement of cultivated plants, which lead to the development of new varieties with more advantageous and superior characters.

1.2.5 Plant Hybridization

Domestication, introduction, acclimatization and selection of plants help to locate the most promising cultivars from the diversity that is available. But, very often, it can be seen that superior and economically important characters are scattered in different cultivars. Hybridization is the technique of bringing superior characters into a single variety by way of cross-pollinating them artificially.

1.2.6 Mutation Breeding

Desirable characters that are scattered in different varieties can be brought together by hybridization. But, sometimes, induction of new variability (new characters and character forms) may become necessary since no cultivars with such traits are available. The most common method that is used to induce new variation is mutation breeding. For the purpose, the seeds or propagules of plants are treated with some chemicals or physical agents that are called *mutagens*.

1.2.7 Polyploidy Breeding

Usually, plants and animals carry chromosomes in pairs in their somatic cells. But, in some cases, more than two sets (multiple sets) of chromosomes (three sets, four sets etc.) can be seen. Such organisms are called *polyploids* and the condition is called *polyploidy*. In the case of some cultivated plants, polyploids show superiority in characters. Breeding of such varieties is called *polyploidy breeding*.

1.2.8 Biotechnological Breeding

Biotechnology is the latest branch of biology that makes use of enzymes as tools to accelerate or manipulate biochemical pathways so as to generate new goods and services based on life and biomolecules. *In vitro* culture technology, marker assisted selection, somatic hybridization, transgenesis and bioreactor technology are the major tools of biotechnology that are useful in plant breeding.

CHAPTER 2

Biological Foundations of Plant Breeding

The science of plant breeding has been developed on the theoretical knowledge of heredity, the impact of environment on characters, systems of reproduction of plants and the science of statistics. The above knowledge constitutes the theoretical background based on which appropriate techniques of breeding for different types of crop plants are developed.

2.1 HEREDITY AND VARIATIONS

Heredity is the phenomenon of inheritance of biological characters of organisms from parental generations to its progeny. In this way, characters of organisms including cultivated plants show a continuum from generation to generation. This continuum of characters is maintained through the transmission of hereditary factors called genes. However, variations that are heritable constitute the raw material for the origin and development of genetic diversity in any population. Heritable variations are caused by changes at the level of genes through mutations and the redistributions and combinations of such changes.

2.1.1 Hereditary Factors

The role of hereditary factors in the control of characters, their nature of action, their transmission from parents to progeny, etc., were first proposed by the Austrian monk, Gregor Johann Mendel. His proposals were based on the results of his experiments in garden pea (*Pisum sativum*). Gregor Mendel conducted hybridization experiments to study the behaviour of hybrid generations. For the purpose, he studied seven pairs of characters in *Pisum sativum* (Table 2.1).

Table 2.1 Characters of *Pisum sativum* studied by Gregor Mendel

Types of character	*Characters*	*Character forms*	*Factors involved*
I. Plant characters	Height of stem	Tall Dwarf	TT, Tt tt
	Flower position	Axial Terminal	AA, Aa aa
II. Pod characters	Pod shape	Inflated Constricted	II, Ii ii
	Pod colour	Green Yellow	GG, Gg gg
III. Seed characters	Seed coat colour	Grey White	WW, Ww ww
	Seed shape	Round Wrinkled	RR, Rr rr
	Cotyledon colour	Yellow Green	YY, Yy yy

Mendel mainly conducted monohybrid and dihybrid crosses in *Pisum sativum*. In monohybrid crosses, he selected one character pair at a time. From monohybrid crosses, he observed that a biological character was controlled by a pair of factors (alleles). In modern genetics, the term allele is synonymous to gene. He further observed that alleles exist in two forms—the dominant form and the recessive form. Dominant forms are capable of controlling the expression of characters by way of controlling the pathways of particular biochemical reactions. Whereas, recessive forms are usually inactive forms of genes, thus resulting in the expression of the character in the absence of the biochemical reaction. Gregor Mendel observed that since the alleles are seen in pairs in organisms, three combinations are possible: homozygous dominant, heterozygous and homozygous recessive (for example, if '**A**' and '**a**' form a pair of alleles, **AA** is the homozygous dominant form, **Aa** is the heterozygous form and **aa** is the homozygous recessive form). Presently, it is known that a biological character is controlled by a pair of factors in many cases. In other cases, where more number of factors are involved, the behaviour of such factors can be analyzed at the level of pairs.

Gregor Johann Mendel: The Austrian monk, Gregor Johann Mendel (1822–1884), discovered the laws of heredity—the basis of modern genetics. Hence, Mendel is called The Father of Genetics. He was born and brought up in an agricultural village in the empire of Austria. In 1843, he joined the Augustinian Monastery at Brunn, which is presently situated in the Czech Republic. In 1848, he completed his theological studies and was ordained as priest. In 1850, he joined the University of Vienna to study science and mathematics. In 1854, he joined the Brunn Modern School as a teacher of physics and natural science. During 1856 to 1866, he published the results of his hybridization experiments conducted on *Pisum sativum* (garden pea).

2.2 INHERITANCE OF CHARACTERS

Biological characters are inherited from the parents to the progeny through hereditary factors that determine the character. Now it is known that the hereditary factors (genes) are present in chromosomes. Chromosomes are certain stainable bodies which show phases of coiling and uncoiling in the course of cell cycle. At the time of sexual reproduction, a special type of cell division known as *meiosis* (also called reduction division) takes place. During meiosis, the chromosomes that are present in pairs (one factor each for each character per chromosome) get separated and move to different gametes. Thus, the gametes carry only single factor. The factors regain the paired nature by the fusion of gametes at the time of fertilization. In this way, factors are inherited from both the parents equally and thus characters also. However, the progeny expresses the dominant form of the character in the homozygous dominant and heterozygous conditions and the recessive form of the character in the homozygous recessive condition.

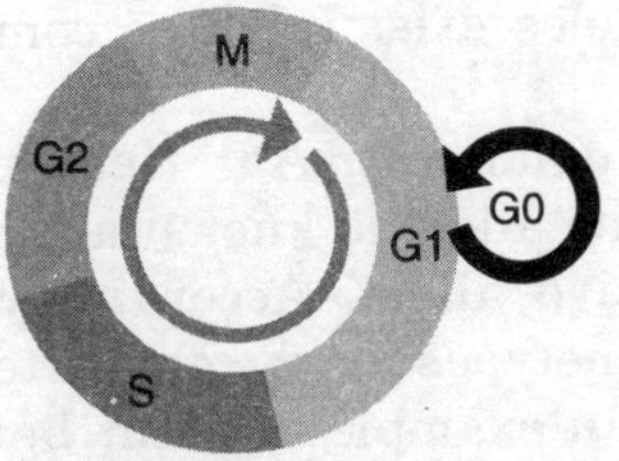

Cell Cycle: G1: First gap phase; **S:** Synthesis phase; **G2:** Second gap phase; **M:** Division phase; **G0:** Gap phase in the case of cells which do not undergo division immediately.

Cell cycle is the life-cycle of a cell in which it grows, synthesizes the chemicals necessary for a new division phase and then divides to give rise to daughter cells identical to the parental cell. Cell cycle is divided into four stages: The first gap (G1) stage that immediately follows the division stage. In the case of cells which do not divide immediately, the cells go to a prolonged resting period called G0. In the case of cells that immediately enter the next division cycle, the G1 phase is followed by the S phase (synthesis phase) in which the DNA and other chemical molecules needed for the next division cycle are synthesized. After the synthesis phase, the cell enters a second gap period known as G2 phase. G2 is then followed by the division stage.

Two types of cell divisions are seen in higher organisms like plants and animals: (1) Mitosis or equational division and (2) Meiosis or reduction division. Mitosis is associated with the somatic growth of organisms and meiosis is associated with their sexual reproduction. As a result of a mitotic division, two daughter cells with the same number of chromosomes (diploid) as that of the parent cell are formed. Whereas in meiotic division, four daughter cells with half the number of chromosomes (haploid) of the parent cell are produced. These act as gametes, fuse in pairs at the time of fertilization and restore the chromosome number as that of the parental cells.

2.2.1 Different Types of Character Expression

Hereditary characters show different types of expression depending on the nature and number of factors involved in their control.

Dominance

In this type of character expression, two alleles present on the homologous loci of a pair of chromosomes control the expression of the character. The alleles exist in three combinations and express characters as follows:

AA: Homozygous dominant; **Aa**: Heterozygous dominant; **aa**: Recessive. **A** form of the allele is known as the *dominant allele* and **a** form as the recessive allele. Accordingly, there will be three genotypes and two phenotypes for the character. Here we can discuss dominance by taking an example of plant height in *Pisum sativum* (Cross 2.1)

Cross 2.1 Genotypes (genetic configuration) and phenotypes (morphological appearance) in dominance. Example: Plant height in *Pisum sativum.*

Parents:	Tall	×	Dwarf
	TT		tt
Gametes	T		t
F_1 :		Tt	
		Tall	

Genotype	*Phenotype*	*Genotypic ratio in* F2	*Phenotypic ratio in* F2
TT	Tall	1 (Homozygous)	
Tt	Tall	2 (Heterozygous)	3
tt	Dwarf	1 (Homozygous)	1

Codominance

This is the condition in which a character is controlled by a pair of alleles and both the alleles are equally dominant. As a result, a third phenotype which is different from the other two is shown by the heterozygous organism (Cross 2.2).

Cross 2.2 Genotypes and phenotypes in codominance. Example: MN blood group system of man.

Parents:	M group	×	N group
	MM		NN
Gametes	M		N
F_1 :		MN	
		MN group	

Genotype	*Phenotype*	*Genotypic ratio in* F_2	*Phenotypic ratio in* F_2
MM	M group blood	1	1
MN	MN group blood	2	2
NN	N group blood	1	1

Incomplete dominance

Here, the dominant allele is only incompletely dominant over the recessive allele. As a result, the heterozygous genotype produces a third phenotype in which the dominant form of the character is only incompletely expressed (Cross 2.3).

Cross 2.3 Genotypes and phenotypes in incomplete dominance. Example: Inheritance of flower colour in *Mirabilis jalapa*.

Parents:	Red	×	White
	RR		rr
Gametes	R		r
F_1 :		Rr	
		Pink	

Genotype	*Phenotype*	*Genotypic ratio in* F_2	*Phenotypic ratio in* F_2
RR	Red	1 (Homozygous)	1
Rr	Pink	2 (Heterozygous)	2
rr	White	1 (Homozygous)	1

2.2.2 Different Types of Control of Characters in Organisms

Most of the biological characters are controlled by hereditary factors present inside the nucleus of the cells. Some of them are controlled by hereditary factors that are present inside the cytoplasm. Characters controlled by factors present inside the nuclei show Mendelian pattern of inheritance, whereas cytoplasmic characters show non-Mendelian pattern of inheritance. Mendelian characters can be classified into two, based on their nature and genetic control—discontinuous and continuous characters.

Discontinuous characters

Discontinuous characters are otherwise called *qualitative characters*. They can be observed by the visual or subjective assessment of the quality. For example, flower colour (Red or White, Red, Pink and White, etc.). No intermediate forms are usually found. The character is expressed as a result of the occurrence or non-occurrence of a biochemical reaction. The character and the factors controlling it show discrete distribution (binomial or polynomial). Usually, only presence or absence of the hereditary factor determines the phenotype of the organism. Plant height and flower colour in *Pisum sativum*, pigmentation in rice etc. are the example of qualitative characters.

Continuous characters

These are the also called *quantitative characters*. They are usually observed by measurement (for example, plant height, fruit weight etc). Intermediate forms can be observed. The character is controlled by a number of hereditary factors, each factor contributing its share towards the expression of the character. The character shows continuous distribution (normal distribution). It is not the presence or absence of alleles, but the member of dominant alleles present, which determines the nature of the characters. Plant height in most of the crop plants, fruit size and number in crop plants etc. are the examples of quantitative characters.

Cytoplasmic characters

These are the characters controlled by hereditary factors present in the cytoplasm. A typical example is cytoplasmic male sterility. This type of characters do not show Mendelian pattern of inheritance. They are inherited from the maternal parent to the progeny through the cytoplasm. Cytoplasmic characters are usually inherited maternally, i.e., from the mother to the progeny. Usually, the cytoplasmic DNA responsible for the character is transferred from the mother to the progeny through the egg cytoplasm (for example, inheritance of plastid colour in plants, inheritance of cytoplasmic male sterility in plants etc.). In some other cases, transcribed information in the form of mRNA is being transferred through the maternal cytoplasm. These particles are known as informosomes (for example, inheritance of shell coiling in snails).

2.2.3 Mechanisms of Inheritance of Characters

Independent assortment

Genes are located on chromosomes. The number of chromosomes in an organism is far below the number of genes. For example, in the rice plant there are thousands of genes but only twelve chromosomes in the haploid set. Hence, it is clear that several genes are present on the same chromosome. It shows that there are two types of inheritance for chromosomal genes—inheritance of genes located on different chromosomes and inheritance of genes located on the same chromosome. Genes located on different chromosomes show independent assortment and genes located on the same chromosome show joint segregation.

Independent assortment is the phenomenon in which a character is inherited from the parents to the progeny independent of all other characters. This type of inheritance is possible only when the genes for the characters are located on different chromosomes. The formation of gametes and the union of gametes in this case is according to the possibilities under probability distribution models.

Independent assortment can be easily explained in plants. Plant height and flower colour in *Pisum sativum* are controlled by two different sets of alleles that are located on different chromosomes and hence they are the examples of independent assortment (Cross 2.4). The alleles of the two characters remain together in a hybrid (F_1) but the alleles move to different gametes at random at the time of gamete formation and they reunite at random at the time of fertilization. Any type of male gamete has got equal chance to get united with any type of female gamete. This phenomenon in which there is non-preferential and random union of alleles of different characters since they are located on different chromosomes is called *independent assortment*.

Cross 2.4 Example for independent assortment from *Pisum sativum* (Tall is dominant over Dwarf and Red is dominant over White)

Parents:	Tall Red	×	Dwarf White	
	TTRR	×	ttrr	
Gametes:	TR		tr	
F_1:		TtRr		
F_2				
Gametes	TR	Tr	tR	tr
TR	TTRR Tall Red	TTRr Tall Red	TtRR Tall Red	TtRr Tall Red
Tr	TTRr Tall Red	TTrr Tall White	TtRr Tall Red	Ttrr Tall White
tR	TtRR Tall Red	TtRr Tall Red	ttRR Dwarf Red	ttRr Dwarf Red
tr	TtRr Tall Red	Ttrr Tall White	ttRr Dwarf Red	ttrr Dwarf White
Genotypic ratio	1 TTRR:2TTRr:2 TtRR:4TtRr:1 TTrr:2Ttrr: 1ttRR: 2ttRr:1ttrr			
Phenotypic ratio	9 Tall Red:3 Tall White:3 Dwarf Red: 1 Dwarf White			

The characters studied by Mendel: Gregor Johann Mendel was lucky enough to find seven pairs of characters in *Pisum sativum* that are located on different chromosomes and show independent assortment. As a result, he was able to formulate three principles—principle of dominance, principle of segregation and principle of independent assortment. Later, the scientist named Correns formulated the first law of Mendelism—the law of segregation and the second law of Mendelism—the law of independent assortment.

As per the law of segregation, the contrasting forms of a factor pair remain together in a monohybrid without mixing up with each other, get distributed in the different gametes randomly at the time of gamete formation and again reunite at random at the time of fertilization. Law of segregation was developed from the findings of monohybrid crosses.

As per the law of independent assortment, the contrasting forms of the factor pairs of two characters remain together in a dihybrid without mixing up with each other, get distributed to different gametes in random combinations of the allelomorphs and the allelomorphs assort independently at the time of union of gametes by fertilization. Law of independent assortment was developed from the findings of dihybrid crosses.

Joint segregation

We know that thousands of genes are present in each and every organism and these genes are located on a lesser number of chromosomes. There is no doubt that several genes are located on the same chromosome. Such genes cannot show independent assortment as in the true sense, genes are not inherited, but chromosomes. The genes located on a particular chromosome get inherited together to the progeny. This type of inheritance is called joint segregation. However, a few recombinants are produced as a result of the genetic recombination that takes place at the time of crossing over.

Maternal inheritance

Maternal inheritance is the inheritance of factors from mother to the progeny. This takes place in the case of characters controlled by plasmagenes (cytoplasmic genes).

2.3 CHARACTER VARIATIONS

Even though the mechanism of inheritance ensures the inheritance of characters from parents to the progeny without change through hereditary factors, very often variations are shown by the progeny. The reasons for the occurrence of such variations are many. The most important reason is the occurrence of mutations. Sexual reproduction is capable of inducing variations in the progeny by the reorientation of genes through crossing over resulting in genetic recombination and also through the union of gametes originating from different parents.

2.3.1 Nature and Types of Mutations

Mutations are mainly classified into two types based on the reason for occurrence. Mutations occur in nature in a low frequency without any evident reason. Such mutations are known as *spontaneous mutations*. Mutations can be induced with the help of certain agents known as *mutagenic agents* or *mutagens*. Mutagens may be physical or chemical in nature. The major physical mutagens include different types of non-ionizing and ionizing radiations whereas the major chemical mutagens include chemicals like alkylating agents, base analogs, certain dyes, etc.

Mutations: Mutations are heritable changes of characters which are originated by changes in the chemistry of hereditary factors. Hereditary factors (genes) are made of DNA (deoxyribonucleic acid) which is the genetic material. Any change in the nature of the DNA molecule may result in heritable character changes. Such changes may occur in nature spontaneously in a low frequency. The frequency of mutations is increased by certain environmental or man-made factors/agents. Such agents are called *mutagenic agents* or *mutagens*. Mutagens have been generally classified into physical mutagens and chemical mutagens. Physical mutagens include ionizing and non-ionizing (exciting) radiations. Gamma rays and X rays are the examples of ionizing radiations whereas UV rays are the examples of non-ionizing radiations. Both excitation and ionization of the molecules of the genetic material DNA may lead to erroneous replication of it ultimately leading to character variations of different kinds. Chemical mutagens cause alkylation or other changes in the case of the nitrogen bases of the genetic material leading to mutations.

2.3.2 The Role of Environment

Hereditary factors are responsible for the inheritance of characters from parents to progeny. Mutations and sexual recombination are responsible for the occurrence of variations. Environment has also got a very significant role in the expression and inheritance of characters. The heritability of characters is highly controlled by environmental factors. Organisms express optimum levels of characters under ideal environmental conditions.

Environmental influence is even more prominent in the case of characters controlled by factors that show genotypic plasticity. This is a phenomenon which shows differential expression of the same character under different environmental conditions.

2.4 SYSTEMS OF REPRODUCTION IN PLANTS

Systems of reproduction in crop plants can be classified into sexual, pseudosexual and asexual types.

2.4.1 Sexual Reproduction

Sexual reproduction in crop plants involves the development of flowers, formation of the floral whorls, development of megaspores and microspores and the formation of egg and male gametes.

Structure of flowers

Flowers are the modified shoots that perform the function of sexual reproduction. They develop usually at the end of the vegetative period of growth of the plants. Usually flowers are borne solitary or in clusters known as *inflorescences*. A flower usually consists of four whorls—calyx, corolla, androecium and gynoecium. Calyx and corolla are accessory floral whorls that provide protection to the other two whorls—the androecium and gynoecium (Figure 2.1). Androecium is the whorl of male reproductive structures and gynoecium is the whorl of female reproductive structures. Androecium consists of

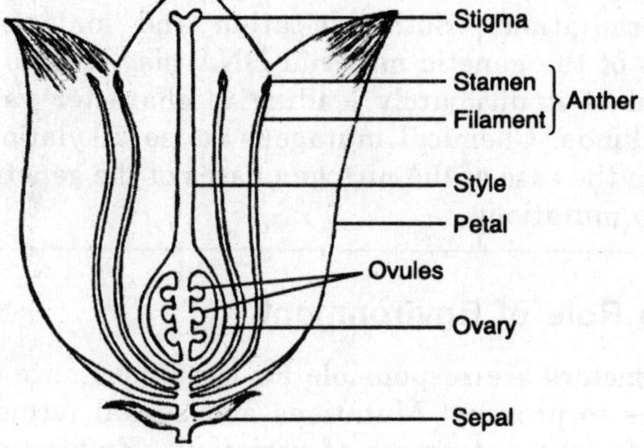

Figure 2.1 Parts of a typical flower.

stamens whereas gynoecium consists of carpels. Each stamen has got a filament and an anther. The microspore mother cells are produced inside the anthers. Each carpel consists of an ovary, style and stigma. Certain structures known as ovules are present inside the ovary. The megaspore mother cells are produced inside the ovules.

Development of microspores and male gametes

The microspore mother cells develop inside the anthers. They are diploid with 2n number of chromosomes. They undergo meiosis (reduction division) and give rise to four haploid microspores per mother cell. Each microspore nucleus further undergoes a mitotic division and forms two haploid nuclei in each cell. One of the nuclei is the generative nucleus and the other is the vegetative nucleus or tube nucleus. This single celled structure with two haploid nuclei represents the *male gametophyte* of plants. The generative nucleus undergoes one meiotic division without cytokinesis to form two sperm nuclei or male gametes.

Development of megaspores and egg

Megaspores develop from a megaspore mother cell situated in the germinal tissue of the ovule. The megaspore mother cell undergoes meiosis and produces four megaspores; three of which degenerate and the remaining one develops into the embryo sac. The embryo sac is surrounded by the maternal tissues of the ovary. The haploid nucleus of the functional megaspores (embryo sac) divides mitotically for three times consecutively, without intervening cytokinesis, resulting in the formation of eight haploid nuclei.

Three nuclei of the embryo sac get oriented near the micropylar end of the embryo sac forming a central egg and two *synergids*. Three other nuclei move to the opposite end of the embryo sac and form *antipodal cells*. The remaining two nuclei seen at the centre of the embryo sac are called *polar nuclei*. The two polar nuclei later unite to form a single diploid nucleus known as *secondary nucleus*. The eight celled mature embryo sac represents the *female gametophyte* of the plants (Figure 2.2).

Pollination and fertilization

Pollination is the transfer of pollen grains from anther to the stigma of the gynoecium. Pollination may be of two types: autogamy or self-pollination and allogamy or cross-pollination. In autogamy, pollen grains of the same flower pollinate the stigma of the flower whereas in allogamy, pollen grains from other plants of the same species pollinate the stigma of a flower.

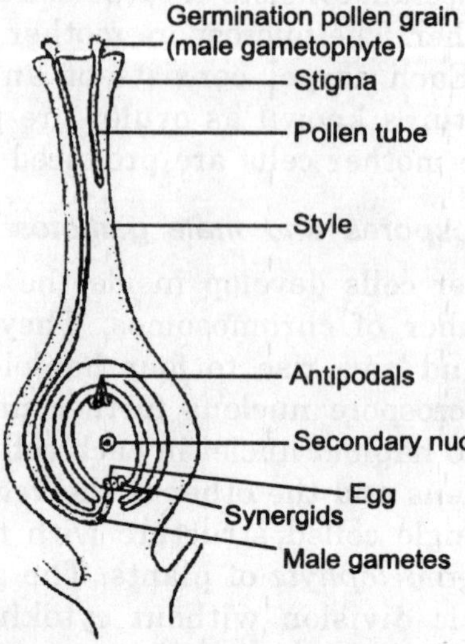

Figure 2.2 Male gametophyte and female gametophyte.

Fertilization involves the fusion of one of the male gametes with the egg and the other with the secondary nucleus both located inside the embryo sac. The fertilized egg develops into the embryo and the secondary nucleus into the endosperm. After pollination, the pollen grain germinates on the stigmatic surface and forms a pollen tube. The pollen tube grows downwards penetrating the style tissue and finally it reaches the ovary, enters one of the ovules and fertilizes it with the male gametes. The fertilized ovules develop into the seeds and the ovary develops into a fruit.

Fruit and seed development

The fertilized ovules develop into seeds. Seeds contain the embryo and the endosperm. A fully grown embryo consists of an elongated embryonal axis and one or two cotyledons attached to it. The ovary develops into a fruit after fertilization.

2.4.2 Pseudosexual Reproduction

Pseudosexual reproduction is a type of reproduction that omits meiosis or fertilization or both. Seed development is without fertilization. Embryo may be either derived from the meiotic division of the megaspore mother cell or from a somatic cell. The process in which embryo develops without the process of fertilization is called *apomixis*. Apomixis is known as diploid apomixis when a diploid embryo is formed and haploid apomixis when a haploid embryo is formed.

Diploid apomixis

This is the type of apomixis in which the embryo formed is diploid. Diploid apomixis may be of different types:

(a) Adventive embryony: When the embryo develops directly from a somatic cell of the nucellus (the mass of tissue present inside the ovule).

(b) Apospory: When the embryo develops from an unreduced egg.

(c) Diplospory: When the embryo develops from a megaspore mother cell mitotically.

Haploid apomixis

This is the type of apomixis in which the embryo formed is haploid. The following are the major types:

(a) Parthenogenesis: When the embryo develops directly from an unfertilized egg.

(b) Apogamy: When the embryo develops from a haploid nucleus other than egg.

2.4.3 Asexual Reproduction

This is the type of reproduction in which the processes of meiosis and fertilization are not involved and reproduction takes place through specialized propagules like bulb, corm, tuber, rhizome, sucker, tiller or runner or through stem cuttings, root cuttings, leaf cuttings, etc. (Table 2.2). Techniques like grafting, budding and layering are also the means of artificial propagation in cultivated plants. Micropropagation through tissue culture is nowadays widely

Table 2.2 Techniques of clonal propagation

Type	*Examples*
Bulb	Onion, Garlic
Corm	Amorphophallus, Gladiolus
Tuber	Potato
Rhizome	Ginger, Turmeric
Tiller	Sugar cane
Runner	Strawberry
Root cutting	Apple, Orange
Stem cutting	Tapioca
Grafting	Cashew, Coffee
Budding	Rubber, Rose
Layering	Fruit trees, Ornamentals
Tissue culture	Many crop plants, Ornamental plants and Medicinal plants

employed to produce bulk quantity of planting materials in the case of a number of crop plants. Different types of vegetative propagation are generally called clonal propagation and the offsprings produced by this method from a single plant are identical. These identical offsprings are called *clones*.

Grafting

Grafting is the art of joining a portion of a plant with buds (scion) on the root stock of another plant. Usually grafting is done so as to attach the growing tip of an improved variety to the root system of a resistant and sturdy variety. Grafting is a very successful method of vegetative propagation in many crop plants like apple, citrus, guava and mango. Different types of grafting are practised in different crops (Table 2.3).

Table 2.3 Types of grafting

1. Splice or whip grafting	Both stock and scion will be of equal diameter. About one-year-old rooted plants are used as root stock. The stock is cut down to a height of about 25 cm above the soil level. A slanting cut is made using a sharp knife. The scion is collected and cut across its lower end, similar to that of the stock. The cut on the root stock should be upward and on the scion downward. The stock and scion are then placed together and held tightly by tying with grafting tape or polythene tape. The wrapping material is removed when the cut surfaces unite.
2. Whip and tongue grafting	Whip and tongue grafting is similar to splice grafting, but a tongue is added to the cut surface for better fitting of the stock and the scion. The stock and scion should be of same diameter.
3. Side grafting	This is the insertion of the scion base under the bark of the stock.
4. Cleft grafting (Figure 2.3)	In this method, the scion is inserted into a vertical slit made in the stub of the plant used as root stock. This method of grafting is commonly used in old trees. Thorough and complete waxing is done to prevent wilting of the scion.
5. Bark grafting	Here, instead of making a slit in the stub, a portion of the bark is split open and the scion is inserted into it.

(Contd.)

Table 2.3 Types of grafting (*Contd.*)

6. Approach grafting (Figure 2.4)	In this technique, two plants (stock and scion) unite together while both of them are on their own root systems. Pot-grown plants or suckers can be effectively used for the purpose. After the healing of the wounds, the top of the stock above the graft and the base of the scion below the graft joint are removed gradually.
7. Top working	In perennial woody plants, existing plants can be improved by grafting scions of superior varieties on them. This technique is called top working. Any method of grafting can be used for top working.
8. Root grafting	This is the method of grafting a shoot or stem (scion) of one plant onto a section of root (stock) of another plant.
9. Epicotyl grafting	In this method, young seedlings (7–10 days after germination) are used as root stock. Cuttings of the same age or higher age can be used as scion. Splice or cleft methods of grafting can be used.
10. Button grafting	This is the type of epicotyl grafting in which seedlings in the button stage are used as scion.

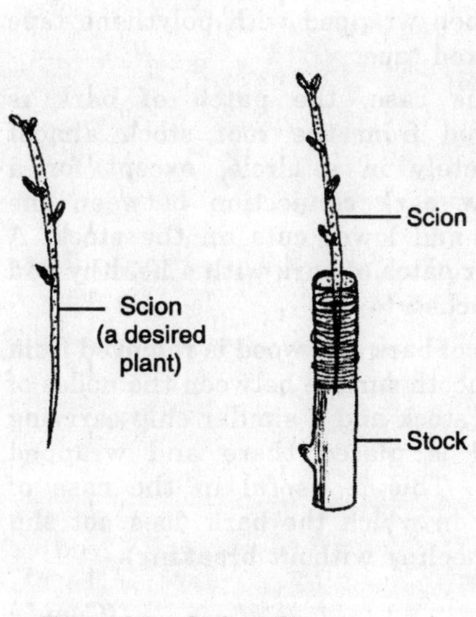

Figure 2.3 Cleft grafting.

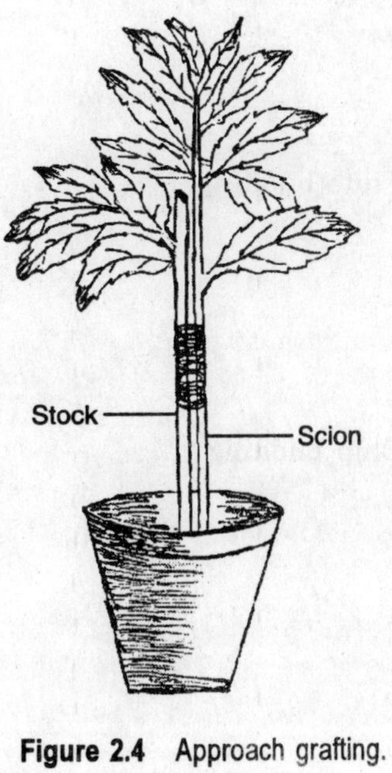

Figure 2.4 Approach grafting.

Budding

This can also be called *bud grafting*. In this method, a single vegetative bud is taken as the scion and it is placed or grafted on a root stock. Budding is done on an actively growing root stock. At this stage, the cambial cells divide actively and the bark can be easily separated from the wood. This stage is called the slipping stage of growth. Different types of budding have been discussed in Table 2.4.

Table 2.4 Types of budding

1. T-budding or shield budding (Figure 2.5)	A T-like cut is made in the bark of the root stock, which is about one year old. The bark in the cut region is slightly separated and the bud is inserted into it. The bud is taken from the scion plant in the form of a shield with the bark region attached to it. The budded region is protected by wrapping with a polythene tape or waxed budding tape. During wrapping, care should be taken to see that the bud remains exposed and all the cut portions are fully covered.
2. Inverted T-budding	Here, the cut in the root stock is made in the form of inverted T. This type of budding is useful in areas with heavy rainfall.
3. Patch budding	A rectangular patch of bark is removed from the root stock and a similar patch with a bud is placed in the cut region and then wrapped with polythene tape or waxed tape.
4. Flute budding	In this case, the patch of bark is removed from the root stock almost completely in a circle, except for a narrow bark connection between the upper and lower cuts on the stock. A similar patch of bark with a healthy bud is attached to it.
5. Chip budding	A chip of bark and wood is removed from the smooth surface between the nodes of a root stock and a similar chip carrying a bud is placed there and wrapped firmly. This is useful in the case of plants in which the bark does not slip well (peeling without breaking).

(Contd.)

Table 2.4 Types of budding (*Contd.*)

6. Skin budding	This involves the removal of the upper side of the bark of the root stock and attaching a similar piece carrying a bud there.
7. Micro-budding	This is a type of T-budding in which a very small shield with bud is used.
8. Top budding	This is best suited to the plants in which budding is done in the place of grafting. This is also practised to rejuvenate old trees.

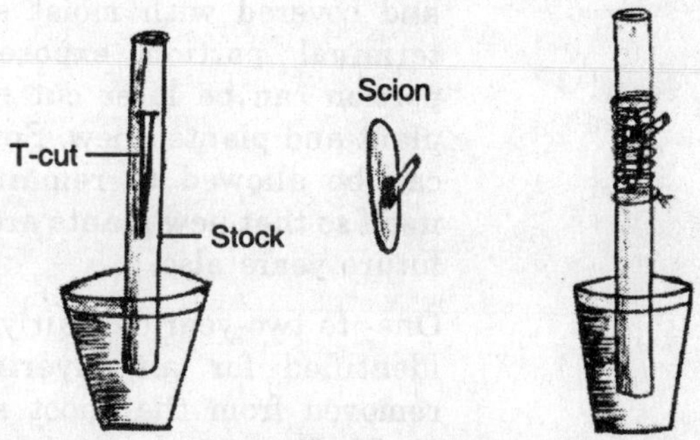

Figure 2.5 T-budding.

Layering

Layering is an age-old technique used to propagate woody plants. This is an improved method of propagation by stem cutting in which stem is induced to root while it is still attached to the mother plant. Many plants produce natural layers. A number of plants can be propagated by means of layers (for example, guava, citrus etc.). When removed from the parent plant, rooted layers are either potted or planted in the nursery bed and maintained carefully until they get established. Different types of layering have been discussed in Table 2.5.

Table 2.5 Types of layering

1. Simple layering	This is a technique in which low, sweeping flexible branches are selected, trimmed, cut to ground level and covered with soil, leaving the tip exposed. The soil around buried region is kept moist to promote rooting.
2. Tip layering	In this case, the growing tips of new shoots are bent and buried in the soil. The curved portion may become etiolated, swollen and

(*Contd.*)

Table 2.5 Types of layering (*Contd.*)

	fleshy and it may give rise to roots within two to three weeks. The shoot tip emerges above the soil surface in the meanwhile.
3. Compound layering (serpentine layering) (Figure 2.6)	In the case of plants with long, slender and flexible stems, the stems can be laid on the ground and alternatively covered and exposed over the entire length. Ringing or girdling of the stems at the regions under the soil will help root growth.
4. Trench layering (continuous layering) (Figure 2.7)	Here, the middle and basal portions of a young stem are placed in a shallow trench and covered with moist soil, leaving the terminal portion exposed. The rooted portion can be later cut from the mother plant and planted new. Some of the shoots can be allowed to remain in the trench itself so that new plants are obtained in the future years also.
5. Air layering (Figure 2.8)	One- to two-year-old fairly long shoots are identified for air layering. Leaves are removed from the shoot and the stem is girdled by removing a ring of bark of about 2–3 cm width, at its base. The girdle interrupts phloem transport, resulting in the accumulation of carbohydrates and hormones above the wound. This promotes fast rooting. The girdled area is then covered with a handful of moist soil mixed with cow dung or compost. This is further covered with sphagnum moss or coir pith and wrapped with a polythene film. The two ends are tied using a twine. The girdled or wrapped region is occasionally moisturized. Rooting takes place within 4–8 weeks. The rooted layers are separated from the parent in two or three stages to reduce the shock of separation. The layers are then planted in pots or nursery beds for establishment.
6. Mound layering (stool layering)	This is a technique of cutting the stem of the plant to a height of 10–15 cm from the ground during the dormant season, allowing new suckers to develop from the buds of the stump. When the new suckers

(*Contd.*)

Table 2.5 Types of layering (*Contd.*)

	attain a height of about 20 cm, they are earthed up by heaping a mound of soil around the stump. As the shoots grow, earthing up is continued until each shoot is buried inside the soil. Afterwards, the rooted shoots are separated from the mother plant and transplanted in nursery beds.

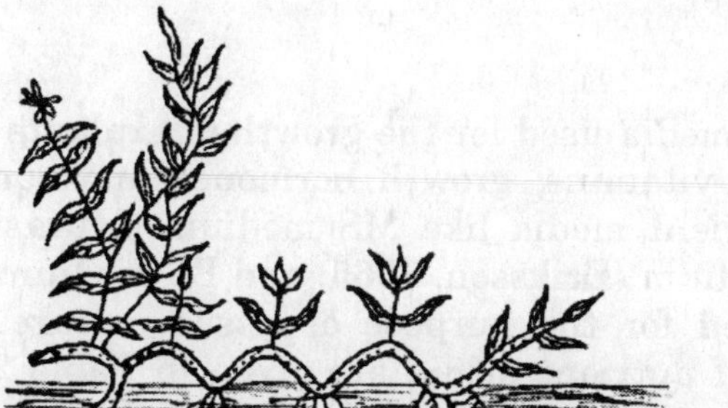

Figure 2.6 Compound layering (serpentine layering).

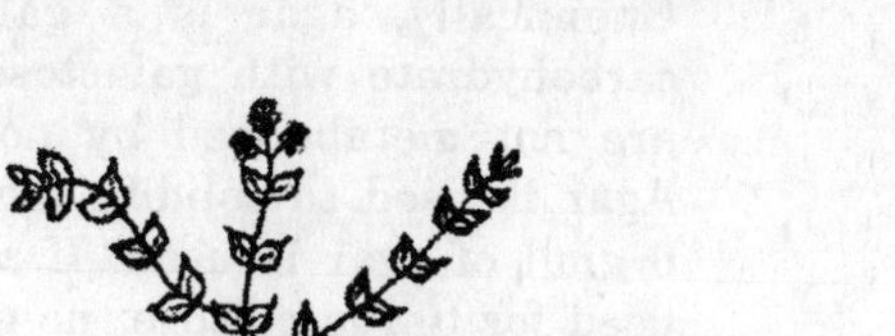

Figure 2.7 Trench layering (continuous layering).

Figure 2.8 Air layering.

2.4.4 Tissue Culture

Tissue culture is the technique of enhancing vegetative plant formation from *in vitro* cultured explants. Direct regeneration of plantlets from cultured plant parts (explants) or regeneration through callus are employed for the purpose. For both the purposes, plant parts are grown in culture media under aseptic conditions. Hormonal proportions are altered so as to obtain callus formation or direct regeneration. The callus or plantlets formed are transferred to hardening media for hardening and acclimatization which is followed by field transfer.

Culture media

The artificial media used for the growth of explants contain sucrose, organic salts, vitamins, growth hormones and some amino acids. Different nutrient media like MS medium (Murashige and Skoog, 1962), ER medium (Eriksson, 1965) and B5 medium (Gamborg *et al.* 1968) are used for the purpose of tissue culture. The important constituents of nutrient media are given in Table 2.6.

Table 2.6 Major components of tissue culture media

1. Agar	Chemically, agar is a galactan, a complex carbohydrate with galactose molecules which are not metabolized by non-ocean microbes. Agar is used to solidify the medium. Usually 6 gm/l of agar is used. If a liquid medium is used for tissue culture, no agar is required.
2. Organic compounds	Organic compounds are used as the source of carbon and energy. Sucrose is usually used in all standard media.
3. Inorganic compounds	Inorganic compounds used in culture media include macronutrients and micronutrients. N, P, K, Ca, Mg and S are the important macronutrients that are used in the preparation of culture media, whereas the micronutrients include Bo, Mo, Cu, Zn, Mg, Fe, etc. A concentrated stock of macro- and micronutrients is prepared and finally added to the medium as and when required. To overcome the problem of solubility, the stock solution of iron is prepared in a chelated form, namely the sodium salt of Ferric Ethylene Diamine Tetra Acetate (EDTA).
4. Growth hormones	Hormones such as cytokinins, auxins and gibberellins are used to regulate growth in tissue culture. Cytokinins promote cell division

(Contd.)

Table 2.6 Major components of tissue culture media (*Contd.*)

	and regulate growth. The most widely used cytokines are adenine, kinetin, zeatin and benzyl adenine. Auxins stimulate shoot elong tion. Auxins usually used are indole acetic acid (IAA), naphthalene acetic acid (NAA), and 2,4-dichlorophenoxyacetic acid (2,4-D). Gibberellins are of less importance. However, gibberellic acid (GA3) is used in apical meristem culture.
5. Vitamins	Vitamins regulate the metabolic activities of cells. They are required in minor quantities. Vitamin B_1 (thiamine) is used for all types of tissue culture. Nicotinic acid, riboflavin, pyridoxin, ascorbic acid, biotin and cyanocobalamin are also used in different cases.
6. Amino acids	Amino acids such as L-aspartic acid, L-asperagin, L-glutamic acid, L-glutamine, etc. are used in tissue culture as sources of nitrogen.

Sterilization methods

All materials like glassware, work tables, instruments, nutrient media, explants, etc., used in tissue culture are to be sterilized properly. The common methods of sterilization are given in Table 2.7.

Table 2.7 Common methods of sterilization

1. Use of chemicals	Chemicals such as Chromic acid, Mercuric chloride (0.1%), Sodium hypochlorite (0.5%), Calcium hypochlorite (0.5%) and alcohol (70%) are used for the sterilization of glassware, work tables and source materials of explants.
2. Use of oven	A dry heat oven is used to sterilize glassware, metallic instruments, etc. by hot air (200–300°C for 1 hour).
3. Use of autoclaves	Autoclaving is done to sterilize nutrient media, distilled water, etc. with the help of steam (121°C for 30 minutes).
4. Ultra filtration	Vitamins, hormones, etc. are unstable at high temperatures. So, they are sterilized using Seitz filter, Millipore membrane filter, etc.
5. Use of UV light	UV light is used in the incubation chamber to make it germ-free.

The technique of tissue culture

The major steps involved in the method of *in vitro* culture of an explant are surface sterilization of the explant, preparation of the medium, transfer (inoculation) of the explant to the medium, growth of the explant to give rise to a callus, subculturing of the callus, organogenesis and acclimatization and transfer of the plantlet to the field.

(i) Surface sterilization of the explant: The plant part which is used for *in vitro* culture is known as explant. Explants are taken from the most appropriate part of the parent plant (Figure 2.9(1)). Development of a tissue is the result of cell division, cell elongation and cell differentiation. Therefore, explants from young and healthy plants are usually used. The presence of parenchyma is the first consideration, because parenchyma quickly responds to culture conditions. Explants are usually 2–5 mm in size. They are sterilized with 1–2% solution of sodium or calcium hypochlorite or 0.1% solution of mercuric chloride. All the operations in tissue culture should be done under aseptic conditions.

(ii) Preparation of nutrient medium: An appropriate nutrient medium is prepared in advance. The medium may be solidified by using agar (6 gm/l) or may be used as a liquid, depending on the requirement. It is poured to culture vials under aseptic conditions.

(iii) Inoculation: Inoculation is the transfer of explants to culture vials (Figure 2.9(2)). It is done in the inoculation chamber under aseptic conditions.

(iv) Callus growth: The explant, a 2–5 mm sterile segment excised from the appropriate part of the plant, is transferred to the nutrient medium and incubated at 25–28°C in an alternate light and dark regime (usually 12 hrs. each). Nutrient medium, supplemented with auxins, induces cell division. As a result, the upper surface of the explant gets covered by an amorphous mass of loosely arranged thin-walled cells. This mass of tissue is called *callus* (Figure 2.9(3)). It is characterized by abnormal growth and has the potential to produce roots, shoots and embryoids.

Callus formation is controlled by the source of the explant, composition of the medium and environmental factors. Explants obtained from meristematic regions develop more rapidly than those of other tissues.

Usually, callus growth is completed in three stages, namely induction, cell division and cell differentiation. During callus induction, the metabolic rate of the cells increases and cells get packed with metabolic products. This is followed by the rapid division of such cells. In the later stage, cellular differentiation starts and sometimes callus becomes coloured.

(v) Subculturing: When the callus has grown for some days (say 28 days), it is essential to subculture it on a fresh medium. Otherwise, nutrient depletion, accumulation of toxic metabolites and paucity of water are sure to occur, leading to the death of the callus.

(vi) Organogenesis: Organogenesis starts in the callus in response to the stimulation given by the chemicals in the medium. Organogenesis takes place in two stages, namely caulogenesis or shoot initiation and rhizogenesis or root initiation. Both types of organogenesis are controlled by the hormones present in the medium. Skoog and Miller (1957) demonstrated that a high auxin:cytokinin ratio may induce shoot formation. In 1966, Torrey proposed the hypothesis of organogenesis. According to this hypothesis, organogenesis starts with the development of a group of meristesmatic cells called *meristemoids*, which initiate the formation of a primordium. Depending on the factors within the system, this primordium develops into either shoot, root or embryoid. Generally, in dicots, buds develop when cytokinin:auxin ratio is 1:100. On the other hand, callus production will be favoured by an auxin:cytokinin ratio 1:100. The formation of an embryoid from the callus is called *somatic embryogenesis*. If the hormonal conditions are correctly balanced, an entire plantlet can be induced to grow on the culture medium. This process is called *regeneration* [Figure 2.9(4)].

(vii) Direct regeneration: In many plants, subculturing of callus results in undesired variations of clones (somaclonal variations). To avoid this, direct regeneration of the explants into plantlets can be tried. This has been achieved in many plant species by altering the hormonal combination of the culture media.

(viii) Other methods of *in vitro* culturing: Plant parts like anther, pollen grains, embryo and endosperm are also grown *in vitro* for different purposes. *Pollen culture* and *anther culture* are practised to produce haploid plants. *Endosperm culture* is carried out to produce triploids. *In vitro* culture of embryos is very useful in some plant species in which the embryo gets aborted under *in vivo* (natural) conditions. In many orchids embryos are rescued and grown in artificial media to ensure proper development of seeds.

(ix) Acclimatization and transfer to the field: In the last stage, the rooted plantlets are subjected to acclimatization, so that they can easily adjust to the field conditions. The plantlets are taken out from the medium, washed thoroughly in running water to remove the agar and are then put in a low mineral salt medium (LMSM) for

24–48 hours. These plantlets are then transferred to pots containing autoclave sterilized mixture of clay, sand and leaf molds in 1:1:1 proportion. The pot is usually covered with transparent polythene to maintain humidity. It is kept undisturbed for 15–30 days. At this stage, the plant becomes fully acclimatized. Finally, these fully acclimatized plantlets can be transferred to the field.

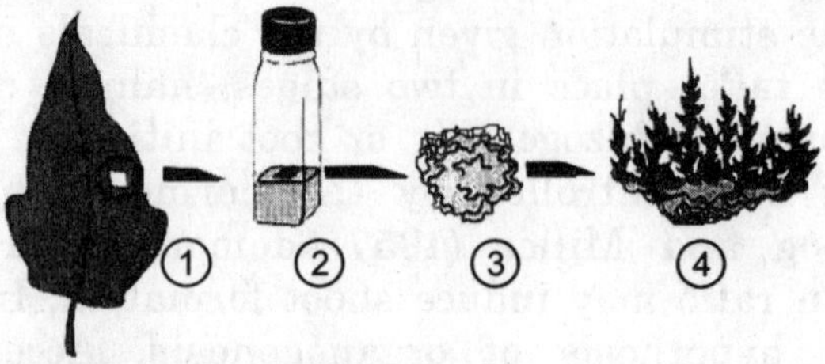

(1) Explant being taken from leaf
(2) Culturing of the explant
(3) Callus formation
(4) Development of plantlets

Figure 2.9 Basic steps of tissue culture.

2.5 MATING SYSTEMS IN SEXUALLY REPRODUCING PLANTS

Three mating systems, *random mating*, *inbreeding* and *outbreeding* can be seen in sexually reproducing populations. Accordingly, the genetic structure of the populations will also be different.

2.5.1 Random Mating

This is the condition in which each male gamete is likely to get equal chance to fertilize a female gamete. This phenomenon is also called *panmixis*. The only factor that acts in the process of fertilization is chance. Under such conditions, the population will be a balanced group of the different genotypes and phenotypes possible, distributed according to theoretical ratios of probability.

2.5.2 Inbreeding

Inbreeding is the phenomenon in which gametes from the same genetic source (gametes from the same genetic structure) get higher chance to mate at the time of fertilization. Many crop plants naturally promote inbreeding (for example, rice, wheat and pulses).

Inbreeding results in the improvement of homozygosity in plant populations. Self-pollination (autogamy) promotes inbreeding. There are several mechanisms in nature that accomplish autogamy. These mechanisms are mentioned below:

Bisexuality

This is the condition in which both the male (androecium) and female (gynoecium) reproductive structures are produced by the same flower. Bisexuality promotes self-pollination. Capsicum, pulses, etc. are the examples of bisexual flowers.

Homogamy

This is the condition in which both the male and female reproductive parts of a bisexual flower mature simultaneously. The androecium and gynoecium mature at the same time to promote self-pollination. In such flowers, mostly pollination takes place before the opening of flower. Capsicum is an example.

Cleistogamy

In some cases the flowers never open and pollination takes place in the bud condition. This is called *cleistogamy*. As a result of this, chance of cross-pollination is eliminated. *Arachis hypogea* is an example.

2.5.3 Outbreeding

Outbreeding is the phenomenon in which gametes from different parental sources get higher chance to unite at the time of fertilization. There are many crop plants which show cross-pollination (allogamy). For example, coconut, maize, sunflower and pearl millet. The following mechanisms are involved to promote cross-pollination in such plants.

Monosexuality

This is the condition in which male and female flowers are separate. Both the types of flowers may be seen on the same plant (monoecy, for example, coconut) or on different plants (dioecy, for example, papaya).

Dichogamy

In many bisexual flowers, androecium and gynoecium mature at different times so as to promote cross-pollination. This phenomenon is called dichogamy. It may be of two types: protandry in which androecium (male reproductive organ) matures first and protogyny (female reproductive organ) in which gynoecium matures first. Both protogynous and protandrous varieties of walnut have been reported.

Heterostyly

This is a condition in which two or more types of styles are seen in some plants. Accordingly, different types of staminal filaments are also seen. In flowers with long styles, short staminal filaments are seen, whereas in flowers with short styles long staminal filaments are present. Primrose (*Primula vulgaris*) shows heterostyly.

Incompatibility

Incompatibility is a phenomenon in which a plant with normal pollen and ovules fails to produce viable seeds through self-pollination. Incompatibility may be of two types—self-incompatibility and cross-incompatibility. If the normal pollen fails to set seed on another plant it is called cross-incompatible and if it is ineffective on the same plant it is called self-incompatible. The incompatibility reaction can occur at the stigma, style or ovary. This phenomenon is genetically controlled, though developmental processes play an important role.

There are two types of incompatibility—gametophytic and sporophytic.

(i) Gametophytic incompatibility: In gametophytic incompatibility the pollen–pistil interaction is governed by the S alleles in the haploid genome of the pollen and diploid genome of the pistil tissue. Gametophytic systems of incompatibility are generally homomorphic. The flowering forms are same on all plants and are related to the capacity of pollen to germinate on the stigma. It is called gametophytic incompatibility because it is determined by the genotype of the gametophytic pollen. *S* locus with several alleles *S*1, *S*2, *S*3, *S*4, ... controls the system. The pistils being sporophytic possess two of these alleles and pollen possesses only one allele. Compatible pollination takes place only when an allele in the pollen is met with different alleles in the pistil. This phenomenon is known to occur in *Trifolium*, *Nicotiana*, *Lycopersicon*, *Solanum*, *Lilium*, etc.

(ii) Sporophytic incompatibility: In sporophytic incompatibility also a single locus *S* system with multiple alleles control the phenomenon. In this system, pollen reaction is premeiotic in action and is determined by the diploid genome of the diploid tissue of the pollen producing sporophyte. Here the phenomenon of dominance is involved. For example, in a plant possessing the genetic constitution *S*1*S*2, all pollen will function as *S*1 if *S*1 is dominant to *S*2 (*S*2 pollen will also show *S*1 reaction on the stigma and style). This type of incompatibility has been considered to be a minor feature in angiosperms, but it is now known in sunflower, cabbage, etc.

Male sterility

Male sterility is also a system to enforce outcrossing. In this case, the pollen is sterile or non-functional. Male sterility is determined by chromosomal factors and/or cytoplasmic factors.

(i) Genetic Male Sterility (GMS): Male sterility in plants is usually controlled by one or two allele loci. Recessive male sterility controlled by a single locus has been seen in some plants. Here, *Ms* allele brings about male fertility and in its absence the plants will act as females.

(ii) Cytoplasmic Male Sterility (CMS): This is a condition in which the presence of a cytoplasmic factor (*S*) determins the male sterility in plants.

(iii) Cytoplasmic Genetic Male Sterility (CGMS): This is a system in which the interaction between genetic and cytoplasmic factors determines male sterility. However, the presence of the dominant chromosomal factor for male fertility invariably causes male fertility. The cytoplasmic factor can cause male sterility in the absence of dominant genetic factors for male fertility.

2.6 BIOMETRICAL GENETICS AND PLANT BREEDING

Biometrical genetics, which is based on the principles and techniques of statistics, is highly utilized in the designing, analysis and interpretation of plant breeding experiments. The genetic structure and structural changes of inbreeding, panmictic and outbreeding populations can be analyzed using biometrical tools. The behaviour of experimental populations and the effectiveness of breeding procedures are analyzed biometrically. Such tools can be used to analyze both qualitative and quantitative characters. Moreover, the designing of plant breeding experiments is based on the principle of statistics.

2.6.1 Factors Involved in an Experiment

Experiments related to the screening of plants are conducted either in net houses or green houses under controlled conditions or in the open field under technical supervision. When a field experiment is designed, many factors like soil heterogeneity, replications, border effect and randomization are taken into consideration.

Replication

The most important factor in the conduct of an experiment is the replication of the treatments. The repeated application of the treatments under investigation is known as *replication*. It is possible to calculate an estimate of error of the experiment from replicated trials. Experimental error gets reduced with the increase in the number of replications. But, in practice, an optimum number is to be fixed for each type of experiments. The number of replications will depend on the degree of accuracy required, availability of land and the quantity of seeds available.

Randomization

Random allocation of treatments to various plots is necessary for an objective comparison between treatments. By randomization, better estimate of the average impact of environmental factors on the outcome becomes possible.

Local control

This is the grouping of homogenous experimental units into blocks and the study of within block variations. The variations within block are caused by soil heterogeneity and other local effects. One such effect is known as *border effect*. It is the harmful or useful effect enjoyed by border lines of the experimental plants. To avoid border effect, separate border rows of plants are grown surrounding the experimental plots.

2.6.2 Experimental Designs

Depending upon the scope and objectives of the experiment, an experimental design which is most suitable for the purpose is selected. The primary intention of designing experiments is to divide the experimental area into homogenous blocks with respect to soil fertility and other ecological factors involved. The aim is to reduce experimental error. Reduction of experimental error depends upon the size, shape and structure of the experimental unit. In plant breeding, mainly three types of designs are employed.

Screening designs

These are the designs used for the preliminary screening of germplasm collected directly from the field. Basically, these designs contain a set of accessions replicated for a convenient number of times and sown in blocks for primary screening.

Evaluation designs

After the preliminary screening, the germplasm is evaluated for desirable characters and for this purpose evaluation designs are used. These comprise complete or incomplete block designs. For small number of entries, complete designs are used and for large number of entries, incomplete designs are used. The following are the complete designs that are used widely.

(i) Completely Randomized Design (CRD): This is the simplest experimental design that can be used for a number of purposes. Here, the experimental field is divided into several plots and the total number of replications (treatments × replications) is allotted at random to each plot simply based on the principle of probability. CRD is appropriate only when the experimental field is homogenous. It is usually preferred in laboratory experiments, green house, or net house studies.

One of the major advantages of CRD is that the number of replications can be varied from treatment to treatment. Because of this, all the available experimental material can be utilized. This provides maximum number of degrees of freedom for the estimation of experimental error.

(ii) Randomized Block Design (RBD): In this case, the experimental field is divided into as many blocks of equal dimensions as there are replications. Each block is then subdivided into as many plots as there are treatments. In each block, each treatment occurs only once (Table 2.8).

Table 2.8 Layout for RBD for four treatments and three replications per treatment

Treatments	*Replications (3 in number)*		
T1	R1	R2	R3
T2	R1	R2	R3
T3	R1	R2	R3
T4	R1	R2	R3

(iii) Latin Square Designs (LSD): These are square designs in which there are as many replications as there are treatments (i.e., 4 × 4, 9 × 9, etc.). It is an improved layout over RBD. Though the treatments are arranged in a random order within each replication, no one treatment can occur more than once in a particular column or row. For randomizing the order of variants, a pack of numbered cards may used (this can be done in the case of RBD also).

RBD is superior to LSD in many ways. RBD is available for a wide range of treatments and there is no restriction on the number of replications. If there is an attack of pests or diseases in one or two blocks, the data from these blocks can be easily omitted without any complication in the analysis. RBD can be conducted in fields of any shape while LSD needs a square plot.

Genetic designs

These are designs that provide estimates of genetic parameters. Three genetic designs, namely North Carolina Design I, North Carolina Design II and North Carolina Design III, and diallele and triallele mating designs come under this category. North Carolina designs are used for segregating populations.

2.6.3 Analysis of Numerical Data

Quantitative data collected from experimental populations are analyzed with the help of different statistical tools so as to reach inferences on different aspects under study. Techniques like study of central tendencies, study of dispersions, tests of significance, analysis of variance, correlation studies, regression studies, factor analysis, cluster analysis, etc. are used to analyze numerical data.

Analysis of central tendencies

Computing central tendencies (averages) is very important to study the general tendency of the behaviour of populations (for example, average yield, average crop duration, etc.). The major measures of central tendencies used for the purpose are following:

(i) Arithmetic mean: The most popular and widely used measure for representing the entire data by one value is arithmetic mean. This is determined by adding all the values of the distribution and dividing the sum total by the number of items.

Depending upon the nature of the data the following formulae are used:

Nature of data	*Formula*
Individual observations (Unorganized data)	Arithmetic Mean $(\overline{X}) = \frac{\Sigma X}{N}$ X represents the observations in the data. N = number of observations.
Discrete frequency table	Arithmetic Mean $(\overline{X}) = \frac{\Sigma fX}{N}$

	X represents the observations in the data; f is the frequency of the corresponding X in the frequency table.
	N = number of observations.
Discrete frequency table	Arithmetic Mean $(\overline{X}) = \frac{\Sigma fm}{N}$
	m represents the mid value of the frequency classes of the data; f is the frequency of the corresponding X in the frequency table.
	N = number of observations.

(ii) Median: The median is defined as the middle value of a distribution. One half of the items of the distribution have got a value of the size of the median or smaller values and the other half have a value of the size of the median or larger values. It is a positional average.

Nature of data	*Formula*
Individual observations (Unorganized data)	Median = size of the $\frac{N+1}{2}$ th item N = number of observations
Discrete frequency table	Median = size of the $\frac{N+1}{2}$ th item N = number of observations
Continuous frequency table	$\text{Median} = L + \frac{N/2 - cf}{f} \times i$ L = Lower limit of the median class N = Number of observations cf = Cumulative frequency of the class preceding the median class f = Frequency of the median class i = Class interval of the median class.

(iii) Mode: Mode is the value in a series of observations that occurs with the greatest frequency.

Nature of data	*Formula*
Individual observations (Unorganized data)	The observation (value) occurring the highest number of times in a distribution is the mode value.
Discrete frequency table	The class with the highest frequency is the mode class. The value of the variable corresponding to the highest frequency is the mode.
Continuous frequency table	$\text{Mode} = L + \left(\dfrac{\Delta 1}{\Delta 1 + \Delta 2}\right) \times i$ L = Lower limit of the modal class Δ1 = The difference between the frequency of the modal class and the frequency of the pre-modal class (ignoring signs) Δ2 = The difference between the frequency of the modal class and the frequency of the post-modal class (ignoring signs) i = Class interval of the modal class.

Measures of dispersion

The measures of central value give us single figures representing the general behaviour of data. But, they give no idea about the variability of the variable being represented in the distribution. It gives no idea about the extreme values of the distribution also. Measures of variation (measures of dispersion) are the statistical values used to know the variations of different values of the variable under study from the central values. The major measures of variation being used in statistical analysis of variability are given below.

(i) Range: Range is the simplest method of studying dispersion of the values of the variable under study. It is the difference between the value of the smallest item and the value of the largest item of the distribution. But, range is a measure of dispersion based on the extreme items of the distribution only.

Range = L–S, where L is the largest value of the variable and S is the smallest value of the variable.

(ii) Variance: It is a statistical measure of variation. It is defined as the mean of squared deviations of the individual values from the mean of the distribution.

Nature of data	*Formula*
Individual observations (Unorganized data)	$\text{Variance} = \frac{\Sigma d^2}{N}$ $d = X - \overline{X}$
Discrete frequency table	$\text{Variance} = \frac{\Sigma fd^2}{N}$ $d = X - \overline{X}$
Continuous frequency table	$\text{Variance} = \frac{\Sigma fd^2}{N}$ $d = \text{mid } X - \overline{X}$

(iii) Standard deviation: Standard deviation is the square root of variance. This is a most widely used measure of dispersion because it is a mathematical measure which shows the extent of variability of each and every value of the variable from its central value.

(iv) Coefficient of variation: It is the most important relative measure of dispersion. It is defined as the sample deviation expressed as percentage of sample mean. In other terms, it is the relative measure of variability in percentage.

$$CV = \frac{SD \times 100}{\text{Mean}}$$

EXAMPLES

I. Individual observations

In a study 30 coconut palms were screened. The following figures were obtained as annual coconut yield. Calculate mean, median and mode of coconut yield. Also calculate variance, standard deviation and the coefficient of variation.

Plant number	*Yield*	*Plant number*	*Yield*	*Plant number*	*Yield*
1	39	11	50	21	48
2	55	12	51	22	49
3	35	13	53	23	31
4	45	14	50	24	33
5	49	15	55	25	50
6	52	16	48	26	51
7	48	17	53	27	51
8	33	18	54	28	48
9	48	19	53	29	48
10	48	20	50	30	51

1. Mean

Arithmetic Mean $(\bar{X}) = \frac{\Sigma X}{N}$

$= \frac{1429}{30} = 47.63$

2. Median

Arrange the data in ascending order. Then median is the value of the $\frac{(N+1)}{2}$th item

1	31
2	33
3	33
4	35
5	39
6	45
7	48
8	48
9	48
10	48
11	48
12	48
13	48
14	49
15	49
16	50
17	50

N = 30

$\frac{N+1}{2} = 15.5$

Hence, the mean of the 15th and 16th items of the variable can be taken as the median value.

15th item = 49

16th item = 50

Median = $\frac{49+50}{2} = 49.5$

18	50
19	50
20	51
21	51
22	51
23	51
24	52
25	53
26	53
27	53
28	54
29	55
30	55

3. Mode

Mode is the most frequent value in the distribution.

Value of the variable	*Frequency*
31	1
33	2
35	1
39	1
45	1
48	7
49	2
50	4
51	4
52	1
53	3
54	1
55	2

The most frequent value of the variable is 48. Hence, 48 is the mode of the distribution. In other words, the highest number of coconut palms yield 48 coconuts.

4. Range

Range = Largest value of the distribution – Smallest value of the distribution.

= 55 – 31 = 24

5. Variance

$$\text{Variance} = \frac{\Sigma d^2}{N}$$

$$d = X - \overline{X}$$

Sl. No.	*Variable* (X)	d	d^2
1	31	–16.63	276.56
2	33	–14.63	214.04
3	33	–14.63	214.04
4	35	–12.63	159.52
5	39	–8.63	74.48
6	45	–2.63	6.92
7	48	0.37	0.14
8	48	0.37	0.14
9	48	0.37	0.14
10	48	0.37	0.14
11	48	0.37	0.14
12	48	0.37	0.14
13	48	0.37	0.14
14	49	1.37	1.88
15	49	1.37	1.88
16	50	2.37	5.62
17	50	2.37	5.62
18	50	2.37	5.62
19	50	2.37	5.62
20	51	3.37	11.36
21	51	3.37	11.36
22	51	3.37	11.36
23	51	3.37	11.36
24	52	4.37	19.10
25	53	5.37	28.84
26	53	5.37	28.84
27	53	5.37	28.84
28	54	6.37	40.58
29	55	7.37	54.32
30	55	7.37	54.32
Σd^2			1273.06

Mean $(\overline{X}) = 47.63$

$$\frac{\Sigma d^2}{N} = \frac{1273.06}{30} = 42.44$$

6. Standard deviation

$$\text{Standard deviation} = \frac{\sqrt{\Sigma}\, d^2}{N}$$

$d = X - \overline{X}$

Sl. No.	*Variable* (X)	d	d^2
1	31	–16.63	276.56
2	33	–14.63	214.04
3	33	–14.63	214.04
4	35	–12.63	159.52
5	39	–8.63	74.48
6	45	–2.63	6.92
7	48	0.37	0.14
8	48	0.37	0.14
9	48	0.37	0.14
10	48	0.37	0.14
11	48	0.37	0.14
12	48	0.37	0.14
13	48	0.37	0.14
14	49	1.37	1.88
15	49	1.37	1.88
16	50	2.37	5.62
17	50	2.37	5.62
18	50	2.37	5.62
19	50	2.37	5.62
20	51	3.37	11.36
21	51	3.37	11.36
22	51	3.37	11.36
23	51	3.37	11.36
24	52	4.37	19.10
25	53	5.37	28.84
26	53	5.37	28.84
27	53	5.37	28.84
28	54	6.37	40.58
29	55	7.37	54.32
30	55	7.37	54.32
$\sum d^2$			1273.06

Mean $\overline{X} = 47.63$

$$\frac{\sum d^2}{N} = \frac{1273.06}{30} = 42.44$$

$$\text{Standard deviation} = \frac{\sqrt{\sum} d^2}{N}$$

= Square root of 42.44 = 6.51

7. Coefficient of variation

$$CV = \frac{SD \times 100}{Mean}$$

$$CV = \frac{6.51 \times 100}{47.63} = 13.67$$

II. Discrete frequency table

Data on the tiller number of rice in an experimental lot is given below. Calculate mean, median, mode, variance, standard deviation and coefficient of variation.

Variable	*Frequency*
10	20
11	20
12	30
13	25
14	15
15	10

1. Mean

$$\text{Mean} = \frac{\Sigma fX}{N}$$

X	f	fX
10	20	200
11	20	220
12	30	360
13	25	325
14	15	210
15	10	150
	$\Sigma f = N = 120$	$\Sigma fX = 1465$

$$\text{Mean} = \frac{\Sigma fX}{N} = 12.21$$

2. Median

$$\text{Median} = \text{Value of the } \frac{N+1}{2}\text{th item.}$$

$$\frac{N+1}{2} = \frac{120+1}{2} = 60.5\text{th item.}$$

The mean of the 60th and 61st values of X can be taken as the median value. To find out median, the cumulative frequencies against each value of X are to be calculated first.

X	f	cf (less than cf)
10	20	20
11	20	40
12	30	70

13	25	95
14	15	110
15	10	120
	$\Sigma f = N = 120$	

The table shows that 60th and 61st values of X lie in the same class where X is 12. Hence, 12 is taken as the median value of the distribution. Median tiller number of the variety of rice is 12.

3. Mode

Mode = the value of X with the highest frequency. In the above distribution, the highest value of f is 30. Hence, the corresponding value of X is the mode value. Therefore, Mode = 12. In other words, 12 is the most frequent tiller number in the case of the rice plants studied.

4. Range

Range = Largest value of X – Smallest value of X
= 15 – 10 = 5

5. Variance

$$\text{Variance} = \frac{\Sigma fd^2}{N}$$

$d = X - \bar{X}$

X	f	d	d^2	fd^2
10	20	–2.21	4.88	97.6
11	20	–1.21	1.46	29.2
12	30	–0.21	0.04	1.2
13	25	0.79	0.62	15.5
14	15	1.79	3.20	48.0
15	10	2.79	7.78	77.8
Mean = 12.21	$\Sigma f = N = 120$			$\Sigma fd^2 = 269.3$

$$\text{Variance} = \frac{269.3}{120} = 2.24$$

6. Standard deviation

$$\text{Standard deviation} = \frac{\sqrt{\Sigma fd^2}}{N}$$

$d = X - \bar{X}$

X	f	d	d^2	fd^2
10	20	–2.21	4.88	97.6
11	20	–1.21	1.46	29.2
12	30	–0.21	0.04	1.2
13	25	0.79	0.62	15.5
14	15	1.79	3.20	48.0
15	10	2.79	7.78	77.8
Mean = 12.21	$\Sigma f = N = 120$			$\Sigma fd^2 = 269.3$

$$\text{Standard deviation} = \frac{\sqrt{269.3}}{120} = 1.50$$

7. Coefficient of variation

$$CV = \frac{\text{Standard Deviation} \times 100}{\text{Mean}}$$

$$= \frac{1.5 \times 100}{12.21} = 12.29$$

III. Continuous frequency table

The following frequency distribution gives data on the number of fruits in a variety of capsicum. Calculate mean, median, mode, variance, standard deviation and coefficient of variation.

Number of fruits	*Number of plants*
30–60	3
60–90	6
90–120	8
120–150	16
150–180	12
180–210	5

1. Mean

$$\text{Mean } (\overline{X}) = \frac{\Sigma fm}{N}$$

X	f	Mid X (m)	fm
30–60	3	45	135
60–90	6	75	450
90–120	8	105	840
120–150	16	135	2160
150–180	12	165	1980
180–210	5	195	975
	$\Sigma f = N = 50$		$\Sigma fm = 6540$

$$\text{Mean} = \frac{6540}{50} = 130.8$$

2. Median

X	f	cf
30–60	3	3
60–90	6	9
90–120	8	17
120–150	**16**	**33**
150–180	12	45
180–210	5	50

Median = Value of N/2th item (in the case of frequency distribution) N = 50; N/2 = 50/2 = 25. From the cf the class in which the 25th item falls can be determined. The class 120–150 is the median class.

$$\text{Median} = L + \frac{N/2 - cf}{f} \times i \qquad 120 + \frac{25 - 17}{16} \times 30 = 120 + 15 = 135.$$

3. Mode

Mode is the value of the variable with the highest frequency. First find out the mode class.

X	f	$\text{Mode} = L + \left(\frac{\Delta 1}{\Delta 1 + \Delta 2}\right) \times i$
30–60	3	
60–90	6	$\text{Mode} = 120 + \frac{8}{12} \times 30$
90–120	8	
120–150	**16**	$= 120 + 20 = 140$
150–180	12	
180–210	5	

4. Range

Range = Largest value of the distribution – Smallest value of the distribution

$$= 210 - 30 = 180$$

5. Variance

$$\text{Variance} = \frac{\Sigma fd^2}{N}$$

d = mid X − $\overline{X}$

X	f	mid X	f mid X	d	d^2	fd^2
30–60	3	45	135	−85.8	7361.64	22084.92
60–90	6	75	450	−55.8	3113.64	18681.84
90–120	8	105	840	−25.8	665.64	5325.12
120–150	16	135	2160	+4.2	17.64	282.24
150–180	12	165	1980	+34.2	1169.64	14035.68
180–210	5	195	975	+64.2	4121.64	321.00
N = Σf = 50						Σfd^2 = 60730.80

$$\text{Variance} = \frac{60730.80}{50} = 1214.62$$

6. Standard deviation

$$\text{Standard deviation} = \frac{\sqrt{\Sigma fd^2}}{N}$$

d = mid X − $\overline{X}$

X	f	mid X	f mid X	d	d^2	fd^2
30–60	3	45	135	−85.8	7361.64	22084.92
60–90	6	75	450	−55.8	3113.64	18681.84
90–120	8	105	840	−25.8	665.64	5325.12
120–150	16	135	2160	+4.2	17.64	282.24
150–180	12	165	1980	+34.2	1169.64	14035.68
180–210	5	195	975	+64.2	4121.64	321.00
N = Σf = 50						Σfd^2 = 60730.80

$$\text{Standard deviation} = \frac{\sqrt{60730.80}}{50} = 34.85.$$

7. Coefficient of variation

$$CV = \frac{\text{Standard Deviation} \times 100}{\text{Mean}} = \frac{34.85 \times 100}{130.8} = 26.64$$

Tests of significance

In plant breeding experiments, the significance of the changes in agronomic and other morphological characters is best analyzed using tests of significances of differences between pre- and post-treatment observations. Tests like Chi-square test and t-test are used for the purpose. The t-test can be used for the comparison of two populations under study. Chi-square test can be used to fit probability/genetic ratios to experimental results and also to test the significance of treatments over control.

(i) t test: t test is used to test the significance of variations between two small populations under study based on their means and standard deviations. This is also used to test the significance of the correlation coefficient. The calculated value of t is compared with the table value of t for the corresponding degree of freedom and the required level of significance. In the case of comparison between two population means, the formula used is given below:

$$t = \frac{\overline{X}_1 - \overline{X}_2}{\sqrt{\frac{s_1^2}{n_1} + \frac{s_2^2}{n_2}}}$$

where $\overline{X}_1$ is the mean of the first population, $\overline{X}_2$ the mean of the second population, s_1 the standard deviation of the first population, s_2 the standard deviation of the second population, n_1 the number of observations in the first population and n_2 the number of observations in the second population.

(ii) χ^2 test: χ^2 test has several uses in test of significance. It can be used to compare two populations based on the variation between observations, it can also be used to test genetic/probability ratios, etc. The formula usually used is given below:

$$\Sigma\chi^2 = \frac{\Sigma(O - E)^2}{E}$$

where, O = Observed values
E = Expected values.

Analysis of variance

Sophisticated analyses like analysis of variance are used in experiments that deal with several characters and several treatments at the same time. Using analysis of variance, variations within samples and variations between samples can be compared and inferences made on the identity of samples. In biological research, analysis of variance can be used to compare the significance of different treatments and also to analyze the statistical significance of variations between different experimental populations/samples under study.

Analysis of variance is otherwise called ANOVA. F is the test statistic calculated in this type of analysis. Here, the treatments are usually more than two in number and each treatment is replicated. Table 2.9 gives an idea of this type of a replicated experiment.

Table 2.9 Analysis of variance

Treatments	*Replications*		
	R1	R2	R3
T1	T1R1	T1R2	T1R3
T2	T2R1	T2R2	T3R3
T3	T3R1	T3R2	T3R3

It is assumed that the variation between replications within a treatment is due to chance (sampling error) and the variation between treatments is due to differential effects of the treatments. The between-treatment variation is higher than within-treatment variation, the variation is considered due to treatment differences. The variance between treatments is compared with variance between replications within a treatment. The test statistic used is F and it is calculated using the formula given below:

$$F = \frac{\text{Between-treatment variation}}{\text{Within-treatment variation}}$$

Following are the steps to calculate the F value:

I. Calculate between-treatment variations:

1. Find out the mean of each treatment.
2. Find out the combined mean of the treatments.
3. Take the deviation of the combined mean from the mean of each sample.
4. Square the deviations.
5. Multiply with the number of replications in the corresponding treatment.
6. Sum the values. The value obtained is called the sum of squares between-treatments (SSB)
7. Divide the sum of squares between-treatments with the treatment degree of freedom (k–1 if k is the number of treatments) to get the mean square (variance) between-treatments.

II. Calculate within-treatment variations:

1. Find out the deviations of the units (replications) of each treatment from the corresponding treatment mean.
2. Square the deviations.
3. Sum the values. Sum the values for all the treatments. The value obtained is called the sum of squares within-treatments (SSW).
4. Divide the sum of squares within-treatments with the degrees of freedom within treatments (n–k if n is the total number of

units in the treatment and k is the number of treatments). The value obtained is the mean square (variance) within treatments.

III. Construct the ANOVA table (Table 2.10)

Table 2.10 ANOVA table

Source of variation	*Sum of squares*	*Degrees of freedom*	*Mean squares*
Between-treatments	SSB	k–1	SSB/k–1
Within-treatments	SSW	n–k	SSW/n–k
Total	SSB + SSW	(k–1)+(n–k) = n–1	

IV. Find out the F value:

$$F = \frac{\text{Variance between-treatments}}{\text{Variance within-treatments}}$$

$$= \frac{\text{Mean square between-treatments}}{\text{Mean square within-treatments}}$$

The calculated value of F is compared with the table value of F for n_1 (between treatment) and n_2 (within treatment) degrees of freedom at the required level of significance. If the table value is lower, calculated F is significant. It shows that the variation between treatments is significant.

Correlation analysis

In practice, we come across a number of variables (characters) that show some related trends of change. This relationship is called *correlation*. Correlation may be positive or negative. Correlation of characters can be calculated as correlation coefficient. Correlation coefficient varies from –1 to +1. When the correlation between two variables is worked out and we get a negative value, the nature of correlation is negative. When we get a positive value, the nature of correlation is positive. In the case of positive correlation, both the variables vary in the same direction. In the case of negative correlation, the variables vary in different directions. If the correlation coefficient is –1, the correlation is perfect negative correlation and if the correlation coefficient is +1, the correlation is perfect positive correlation. If the correlation coefficient is 0, there is no correlation.

Coefficient of correlation is calculated using the following formula:

$$r = \frac{\Sigma xy}{\sqrt{\Sigma x^2 \times \Sigma y^2}}$$

$$x = X - \overline{X}$$

$$y = Y - \overline{Y}$$

Example:

Find out the nature of correlation between the two variables given below:

Plant height (cm) (X):	50	55	60	65	70	75
No. of fruits (Y)	10	12	14	16	18	20

Solution:

X	Y	x	y	xy	x^2	y^2
50	10	–12.5	–5	62.5	156.25	25
55	12	–7.5	–3	22.5	56.25	9
60	14	–2.5	–1	2.5	6.25	1
65	16	2.5	1	2.5	6.25	1
70	18	7.5	3	22.5	56.25	9
75	20	12.5	5	62.5	156.25	25
$\overline{X} = 62.5$	$\overline{Y} = 15$			$\Sigma xy = 175$	$\Sigma x^2 = 437.50$	$\Sigma y^2 = 70$

Correlation coefficient (r) $= \dfrac{\Sigma xy}{\sqrt{\Sigma x^2 \times \Sigma y^2}}$

$$= \frac{175}{20.92 \times 8.37} = \frac{175}{175.1} = 0.999$$

The correlation coefficient = 0.999 = 1. It shows that the two variables show almost perfect positive correlation.

Regression analysis

After establishing the relationship between two variables, the estimation of one variable becomes possible if the other is known, with the help of regression analysis. Regression equations of the two variables under study can be used for the purpose.

Factor analysis

Factor analysis is a very efficient method of data reduction in which data are grouped into different sets of variables. Lead variables can be identified from these groups which can be used as variables and can be targeted in selection and other breeding programmes.

Cluster analysis

This involves grouping of samples based on their affinities. This technique can be used in the study of relationships of samples under study.

CHAPTER

3

Conventional Methods of Plant Breeding

Plant breeding is an art and science that evolved in the hands of farming communities around the world. In the twentieth century, the science of plant breeding developed in the hands of plant breeders specialized in different crop plants. The conventional techniques of plant breeding that are being practised by plant breeders are discussed in this chapter.

3.1 DOMESTICATION

We know that all the crop plants and their relatives originated and evolved in the wild, in the hands of Mother Nature. These plants got domesticated slowly and slowly with the development of agriculture. *Domestication* is the process of bringing wild species under human management. Most of the crop plants were domesticated by prehistoric man. After domestication, the crop plants got changed considerably as compared to their wild forms.

Domestication of wild plants is still being continued and it is likely to continue in future also, since new requirements may necessitate the domestication of new plant species. Even though most of the food crops were domesticated prehistorically, crops like rubber, medicinal plant species etc. were domesticated recently.

Even beverage crops like tea and coffee were domesticated only a few centuries ago. Many latex producing plants are being domesticated recently for their potential ability to yield products like petroleum.

3.1.1 Patterns of Changes under Domestication

Domesticated populations of plants undergo continuous evolution depending upon the pressure exerted by the farmers, plant breeders and the environment. Variability arises in these populations through mutations, recombination, transposition etc. Both natural and artificial selections take place in such populations. Besides, natural and artificial hybridizations also occur in domesticated populations. As a result of these processes, continuous speciation takes place in domesticated populations.

Origin of variability in domesticated populations

A plant population is considerably variable at the time of domestication itself. New variability arises within domesticated populations by mutations, recombination, hybridization etc.

Natural and artificial selection under domestication

As in the case of wild populations, in domesticated populations also, genotypes that are more adapted to the environment give rise to higher number of progeny than that of less adapted genotypes to the prevailing environment. This phenomenon is called *natural selection*. This process leads to the elimination of less adapted genotypes from the gene pool of the populations, thus piloting the process of speciation. Man has also exerted considerable efforts to select favourable genotypes in the process of agriculture and the improvement of cultivated varieties. This is called *artificial selection*. Selection of crop plants for better grain size, fruit size, quality characters, resistance etc. is carried out through artificial selection. This process also has contributed significantly towards the evolution of cultivated plants.

Natural selection is the process of favourable selection or elimination of variations that exist in a population so as either to maintain the population without change, or to give rise to one or more new populations, which in due course get evolved into new species. Accordingly, three types of natural selection are possible:

1. Stabilizing selection: The type of natural selection in which aberrants are eliminated and the population equilibrium is maintained without change.

2. Directional selection: The type of selection in which evolutionary modification gradually progresses in one direction, as a result of which a new species originate from the existing one, replacing it.
3. Diversifying selection: The type of selection in which selection acts in more than one direction, so that more than one species originate from the existing population.

Speciation is the process of accumulation of variations, action of natural selection on the variations and subsequent evolution of new species.

Speciation under domestication

The pattern of speciation of a species gets considerably modified by domestication. Characters related to agronomic properties show higher degrees of variation. The modifications that are usually found associated with speciation under domestication are given in Table 3.1.

Table 3.1 Modifications under domestication

S. No.	*Modifications*
1	Reduction in shattering of fruits
2	Elimination of dormancy
3	Decrease in toxin content
4	Changes in plant type
5	Changes in plant height
6	Reduction of life duration
7	Increase in fruit/grain size
8	Development of polyploids
9	Reduction of variability

3.1.2 Centres of Origin of Cultivated Plants

We have seen that all the present-day cultivated plants originated in the wild. However, these wild ancestral plants were not distributed uniformly through out the world. It was N.I. Vavilov, the great biologist who headed the Institute of Plant Industry of Leningrad (USSR) from 1916 to 1936, who propound that crop plants evolved from wild ancestors in certain areas of great diversity that are called their centres of origin. From these centres of origin, crops were dispersed or transported to other geographical areas in the hands of man.

The main centres of variation of crop plants as described above are called their *primary centre of diversity*. It is the region of true origin. But, in some cases, certain crop species show considerable variability of forms in certain centres, even though they did not originate in these centres. Such centres are known as *secondary centres of diversity*. These are the regions of subsequent spread of a crop.

Eight main centres of origin were originally recognized and proposed by Vavilov (Table 3.2; Figure 3.1).

Table 3.2 Centres of origin of cultivated plants as originally proposed by Vavilov

S. No.	*Name of centre*	*Geographical description*
1	China centre	Mountainous regions of central and western China and the neighbouring low lands.
2	Hindustan centre	The region consisting of Burma, Assam, Malayan Archipelago, Java, Borneo, Sumatra and Philippines.
3	Central Asia centre (Afghanistan centre)	North-west India, Afghanistan, Tajikistan, Uzbekistan and Tianshan.
4	Asia Minor centre (Near East centre or Persian centre)	Asia Minor, Transcaucasia, Iran and the highlands of Turkmenistan.
5	Mediterranean centre	Mediterranean region.
6	Abyssinian centre	Ethiopia and hilly regions of Eritrea.
7	Central American centre (Mexican centre)	South Mexico and Central America.
8	South American centre	Mountainous regions of Peru, Bolivia, Ecuador, Colombia, parts of Chile and Brazil and the whole of Paraguay.

Later, in 1935, Vavilov divided the Hindustan centre into two: Indo-Burma centre and Siam-Malaya-Java centre, whereas South American centre was divided into three: Peru centre, Chile centre and Brazil-Paraguay centre. He also introduced a new centre of origin—US centre, thus making the total number of centres 12 (Table 3.3).

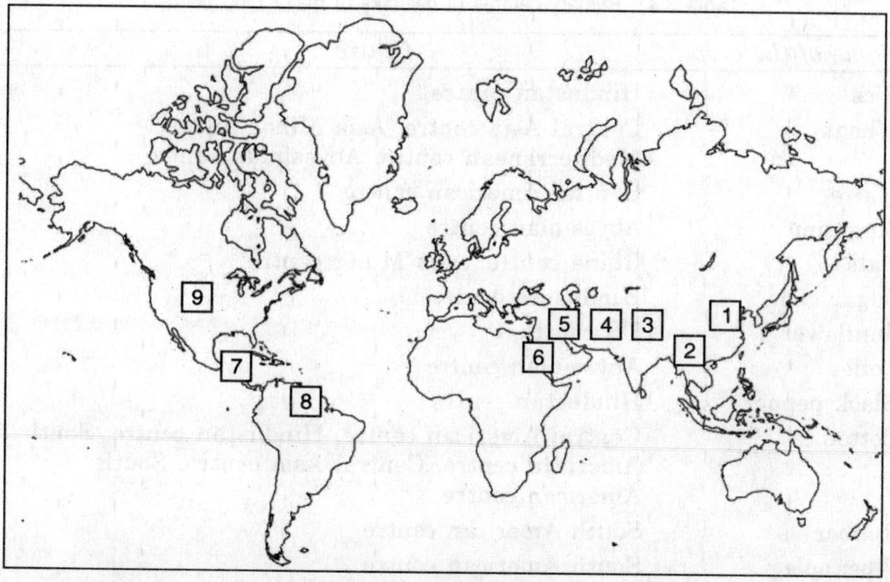

Figure 3.1 Centres of origin of cultivated plants.
1. China centre; 2. Hindustan centre; 3. Central Asia centre; 4. Asia Minor centre; 5. Mediterranean centre; 6. Abyssinian centre; 7. Central American centre; 8. South American centre; 9. US centre

Table 3.3 Twelve centres of origin for cultivated plants (Vavilov in 1935)

S. No.	*Centre of origin*
1	China centre
2	Indo-Burma centre
3	Siam-Malaya-Java centre
4	Central Asia centre
5	Asia Minor centre
6	Mediterranean centre
7	Abyssinian centre
8	Central American centre
9	Peru centre
10	Chile centre
11	Brazil-Paraguay centre
12	US centre

Table 3.4 Primary centres of diversity of major crop plants

Name of the crop	*Centre*
Rice	Hindustan centre
Wheat	Central Asia centre, Asia Minor centre, Mediterranean centre, Abyssinian centre
Maize	Central American centre
Sorghum	Abyssinian centre
Oats	China centre, Asia Minor centre
Sugar cane	Hindustan centre
Sunflower	US centre
Coffee	Abyssinian centre
Black pepper	Hindustan centre
Cotton	Central American centre, Hindustan centre, South American centre, Central Asia centre, South American centre
Rubber	South American centre
Pineapple	South American centre
Peanut	South American centre
Cassava	South American centre
Arrowroot	Central American centre
Papaya	Central American centre
Carrot	Central Asia centre, Asia Minor centre
Apple	Asia Minor centre
Grape	Central Asia centre, Asia Minor centre
Banana	Hindustan centre
Coconut	Hindustan centre

The concept of centres of origin has been however questioned. These centres of origin may be more appropriately called *centres of diversity* (Table 3.4). No doubt that these are areas of maximum diversity of crop species. Within large centres of diversity, micro-centres of diversity can also be observed. Crop evolution is more dynamic in these micro-centres. The micro-centres are important for collection of plant species and varieties for domestication, introduction and *ex situ* conservation.

3.2 PLANT INTRODUCTION

Plant introduction is the process of bringing a plant species or variety to a new country or geographical area where it was not previously grown. Thus, the introduced plant may be a new crop species or variety. Often, such plants are introduced from other countries, continents or geographical areas.

3.2.1 Types of Plant Introduction

Primary introduction

This is the type of introduction in which the introduced species or variety is directly released for cultivation. This type of introduction is rare now. However, rice varieties developed at International Rice Research Institute, Philippines like IR8 and IR20 were directly introduced to India for cultivation, as a part of green revolution.

Secondary introduction

In this case, the introduced variety or species is subjected to further screening experiments via selection or they are used as parents in hybridization programmes.

3.2.2 Agencies of Plant Introduction

In the past, introduction of plants from country to country or continent to continent took place through the hands of travellers, soldiers and kings. For example, mung, mustard, pear, apple and walnut were introduced to different parts of the world, from the Central Asian centre of origin in the medieval period. Botanical gardens have played significant role in plant introduction in the past centuries. Presently, agencies like different International and National Crop Research Institutes, National Departments of Agriculture of different countries, International Plant Genetic Resources Institute (IPGRI), etc. are involved in germplasm collection and effectively co-ordinating plant introduction.

3.2.3 Procedure of Plant Introduction

The procedure of plant introduction consists of procurement, quarantine, cataloguing, evaluation, multiplication and distribution.

Procurement

The material that is required to be introduced from other countries is nowadays procured through National Departments of Agriculture, National Crop Research Institutes or National Bureaus of Plant Genetic Resources. In most of the cases, free and mutual exchange of genetic resources is possible between countries.

Plant propagules like seeds, clonal propagules or cultured tissues can be exchanged. Proper packaging of the material being exchanged is very important. Nowadays, *in vitro* exchange of germplasm is becoming popular.

Quarantine

Quarantine is the screening of the living organisms introduced from foreign sources for the presence of pathogens, pests, weeds etc. Plant materials introduced from such sources are screened and suspected materials are treated appropriately. All the plant materials being introduced should carry an authentic phytosanitary certificate.

Quarantine control is usually exercised by the agencies authorized for introduction, at prescribed ports of entry. Materials contaminated by diseases or pests are destroyed or returned. Materials conforming to quarantine laws are fumigated against diseases and pests. Sometimes they are grown in isolation or quarantine plots for evaluation. Restrictions are imposed on the introduction of plant materials by different countries due to several reasons.

Cataloguing

When a material is introduced, it is given an entry number and information regarding the name of the variety, species, place of origin and major characteristics are recorded.

Evaluation

The newly introduced materials are usually assessed and evaluated by the respective Crop Research Institutes. Resistance to pests and diseases is evaluated under appropriate conditions.

The acclimatization of an introduced variety or species to the environmental conditions of the new geographical area is a serious problem. However, since the introduced lot may contain genetically different plant types, some of them may show higher levels of adaptability to the new environment. Such genotypes are multiplied and distributed widely.

Multiplication and distribution

Superior genotypes are selected, multiplied and made available on commercial scale. They are subjected to field trials and released for cultivation. Some of them are used as parents in hybridization programmes so as to produce improved hybrids.

3.2.4 Genetic Resources and their Conservation

The primary and secondary centres of diversity of crop plants are immensely rich in genetically variable forms of crop plants and their wild relatives. This diversity of genetically different forms in the case of each crop forms its genetic resources. The genetic resource should be explored, evaluated, catalogued and conserved so that no genotype is lost from the treasury of genetic diversity. Such a collection of genetic resources (collected, conserved, catalogued and evaluated exhaustively) is called a *germplasm collection* of a crop. Nowadays several regional, national and international research institutes and agencies are involved in germplasm collection of different crops. Genetic resources can be conserved *in situ* and *ex situ*.

In situ conservation of genetic resources

This refers to the conservation of genetic resources in their natural habitats itself. Wild relatives of plants, weedy relatives, land races etc. are conserved by this method. Such populations keep on evolving in their national environments, broadening their genetic base. Such conserved collections can be called *natural reserves*, *natural parks* or *gene sanctuaries*.

Ex situ conservation of genetic resources

This is the technique of conserving genetic resources under controlled experimental conditions outside their national habitats. *Ex situ* conservation may be either *in vivo* or *in vitro*.

A. *In vivo* conservation: This involves *ex situ* conservation under field conditions. Such germplasm collections can be called field gene banks. In India, *field gene banks* of most of the crops have been conserved by the National Research Institutes concerned. The national field gene bank of rice is maintained by Central Rice Research Institute, Cuttack. Whereas the national field gene banks of pepper and rubber are maintained by Indian Institute of Spices Research, Calicut and Rubber Research Institute of India, Kottayam, respectively.

B. *In vitro* conservation: *In vitro* conservation is the conservation of tissues, cells, genomes and genes under controlled environmental conditions. Tissues and different plant parts are conserved at very low temperature (–196°C) in liquid nitrogen. This technique is called *cryopreservation*. Genes and genomes are conserved in gene/genome/DNA libraries. These types of collections can be called *in vitro* gene banks.

Types of ex situ collections

Four types of *ex situ* germplasm collections are recognized based on the duration and importance of conservation.

A. Base collections: These are long-term collections of germplasm (over 20 years). In base collections, seeds are stored at low moisture levels (3–6%) and zero degree temperature. Whereas other plant parts including cultures are stored under cryopreservation.

B. Active collections: These are collections under medium-term storage (10–15 years). In this type of collection, seeds are stored at the temperature of around 0°C and moisture of 8%. Cultured materials are also conserved under medium-term storage.

C. Working collections: These are collections under short-term storage (3–5 years) and are maintained at 5–10°C temperature with 8–10% moisture content. These are breeders' collections that are utilized for different breeding purposes.

D. Core collections: This includes the entire genetic diversity of a species conserved with minimum replications. This represents a subset of the entire germplasm with all useful characters so that identification of useful entry becomes easy and accessible to breeders.

3.3 SELECTION

All the present-day crop plants are the descendants of the plants that were either domesticated from the wild or introduced from other continents or geographical areas. In the wild condition, their populations had been undergoing continuous speciation through natural selection. But, once domesticated, besides natural selection, artificial selection in the hands of farmers and plant breeders became a very critical force directing the process of speciation in these plant species. In the 20th century, this process evolved to certain scientific techniques based on the genetic structure and nature of reproduction of the population.

3.3.1 Selection in Self-pollinating Crops

Two methods of selection are generally practised in self-pollinating species of crop plants: mass selection and pure line selection.

A. Mass selection

This is the method of improving a crop strain through the selection of a large number of superior plants by visual assessment, pooling of their seeds and developing a new variety from it. Selection is done for easily observable characters like plant height, nature of the panicle, grain or seed size, resistance etc.

The population raised from the selected plants will be more uniform and superior when compared to the original population. However, the population will show considerable variations, especially for quantitative characters. The variations thus produced are sometimes subjected to progeny test.

Progeny test is the technique of assessing the performance of a crop population by analysing the performance of its progeny.

Procedure of mass selection: The procedure of mass selection extends to 7–8 crop seasons in case of annual crops. The procedure involves identification of superior plants from a population and bulking their seeds, preliminary yield trials, multi-location trials, seed multiplication and distribution.

1. Identification of superior plants and collections of seeds

   This is done in the first year of the experiment. A large number of phenotypically similar superior plants are selected and their seeds are bulked.

2. Preliminary yield trials

   This is usually done in the second year or second crop season. The bulked seeds are grown in a preliminary yield trial with standard variation as checks. The variety from which selection is made should also be included as a check in order to determine improvement in characteristics.

3. Multi-location trials

   This is usually done from the third year onwards.

4. Seed multiplication and release

   After multi-location trials, the seeds are subjected to the procedure of variety release and then multiplied and distributed to farmers.

Merits and demerits of mass selection: Varieties developed through mass selection are more widely adapted than pure lines since they are developed from a large number of plants selected for their superior characteristics. The procedure of mass selection is simple and less sophisticated. Mass selection retains considerable degree of genetic variability.

But, the varieties developed through mass selection are non-uniform in comparison to pure lines. The degree of improvement is only limited. Without progeny test, it is not possible to assess the genetic stability of the variety.

B. Pure line selection

Pure line selection involves the development of a new variety from a simple homozygous self-pollinated plant. As a result, all the individuals within a pure line have the same genotype and variations, if any, are due to the environmental influences. In pure line selection, a large number of superior plants are selected from a self-pollinated crop, harvested individually, individual progenies are raised and evaluated and the best progeny is released as a variety. This technique is also known as individual plant selection.

Procedure of pure line selection: The procedure of pure line selection involves individual selection of superior plants and collection of their seeds separately, growing individual plant progenies and rejection of undesirable progenies, preliminary yield trial and rejection of undesirable progenies, multi-location trials and release of the most adaptable and superior progeny as a new variety.

1. Selection of superior plants and collection of seeds

A large number of superior plants are selected in the first year based on visual assessment for characteristics and their seeds are collected separately. Selection is made on the basis of easily observable characters.

2. Growing progeny lines of the selected plants

Progeny lines of individual plants are grown in the second year and inferior progeny lines are rejected. Seeds of each line are collected separately.

3. Preliminary yield trial

Preliminary yield trial of the remaining progenies is carried out in the third year and inferior progenies are rejected. Seeds of each line are collected separately.

4. Replicated yield trials (Multi-location trials)

Replicated yield trials are carried out at several locations and inferior progeny lines are rejected. The lines are screened for resistance and quality trials. This is done from fourth to seventh years.

5. Variety release and seed multiplication

This is done in the eighth year. The best progeny selected from the above is released as a new variety. The seeds are multiplied for distribution.

Characteristics of pure lines: All the plants within a pure line have same genotype, as the parent plants are homozygous and self-fertilizing. Variations, if any, within a pure line are environmental and usually non-heritable. Selection within a pure line is not effective. However, in the course of time, variations produced by mechanical mixtures, mutations and natural hybridization get incorporated into the populations and further cycles of mass selection and pure line selection may become necessary to improve the variety.

Merits and limitations of pure line selection: Pure line selection brings about the maximum possible improvement over the original variety. Pure line selections are extremely uniform in their behaviour. Uniform flowering, maturity etc. are the characteristic adaptations of pure lines. Due to their extreme uniformity, such variations can be easily identified in seed certification programmes.

However, pure line varieties show very narrow range of adaptations to environmental fluctuations. Pure line breeding is a tedious technique and the varieties developed in this way, even though uniform, are shortlived.

3.3.2 Selection in Cross-pollinating Crops

Selection is applied in cross-pollinated species for population improvement. The methods of selection used in the case of cross-pollinated crops are discussed below.

A. Mass selection

Mass selection is the technique in which individual plants are selected on the basis of their performance and the bulked seeds are used to produce the next generation. Since phenotype is the sole criterion of selection in mass selection, its efficiency depends upon the magnitude of heritability. Several modifications of mass selection are also being practised.

Selfed plant mass selection: The available additive and dominant gene variance in a random mating population can be reshuffled through selfing. The selfed progeny rows are selected on the basis of phenotypic performance. They are pollinated by a random bulk of other S_1 plants. The unit of selection and recombination are thus selfed S_1 plants.

Progeny selection: This is mass selection with progeny test. The simplest form of progeny selection is the ear-to-row method. The ear-to-row method can be explained as follows:

A number of plants are selected on the basis of the phenotype in the first season. They are allowed to open pollinate and the seeds from individual plants are harvested separately. In the second season, progeny rows are grown from each selected plant. The progeny rows are evaluated for desirable characters and superior progenies are identified. They are permitted to open pollinate and the progeny is grown in progeny rows again. The cycle is repeated and finally the seeds of superior plants are collected and bulked.

In progeny selection, the selection is made on the basis of progeny test and not on the phenotype of individual plants. This helps in the selection of plants with better genotypes. This is a good tool to increase the yielding ability of open pollinated plants. Inbreeding can be avoided if care is taken to select sufficiently large number of plant progenies. The selection scheme is simple and easy.

The limitation in most of the progeny selection schemes is that there is no control on pollination and the plants are allowed to open pollinate. Besides, selection is based on the maternal parent only which reduces the efficiency of selection.

Line breeding: Line breeding is a variation of progeny selection. In line breeding, one or more cycles of selection are carried out, on the basis of progeny tests. At the end of the selection process the superior progeny lines are composited to produce a new variety.

B. Recurrent selection

According to Hull (1945), recurrent selection is defined as reselection generation after generation with interbreeding of the selected plants to provide for genetic recombination. Different types of recurrent selection that are in practice are described below:

1. Simple recurrent selection: This is a technique of selection in which one original selection cycle and one or more recurrent selection cycles are carried out for population improvement. In the original selections cycle, a number of plants with desirable characters are selected and self-pollinated. Separate progeny rows are grown from the selected plants in the next generation. The progenies are intercrossed in all possible combinations and equal amount of seeds from each is composited to produce the next generation.

In the first recurrent selection cycle, several desirable plants are selected from the composited population on the basis of phenotype and they are self-pollinated. Progeny rows are grown and all possible intercrosses are made by hand. Equal quantity of seeds from all the intercrosses are composited to produce the next generation. Recurrent selection cycles may be repeated if required.

2. Recurrent selection for general combining ability: This is a type of recurrent selection that is used to improve the general combining ability of a population and a heterozygous tester with a broad genetic base is used for this purpose. This method is useful for the genetic improvement of quantitative characters. Selection is made on the basis of testcross performance.

In the first year, superior plants for the desirable character are selected and selfed. These selected plants are also crossed to a heterozygous tester with broad genetic base. The selfed seeds are kept under cold storage. Line × tester crosses are called top crosses.

In the second year, the crossed seeds are grown and the combining ability of the selected plants is evaluated and the plants with good general combining ability are identified.

In the third year, the progenies of the selected plants are grown from their selfed seeds kept in cold storage. Their progenies are intermated in all possible combinations and their crossed seeds are composited to form a new source population for further selection. This completes the original selection cycle. In this way, subsequent cycles (recurrent cycles) can be carried out.

3. Recurrent selection for specific combining ability: This is a type of recurrent selection that is used to improve the specific combining ability of a population for a character using a homozygous tester. This method is used for the genetic improvement of polygenic characters. Selection is made on the basis of testcross performance. An inbred is used as the tester. The selection procedure is same as in the case of recurrent selection for general combining ability.

4. Reciprocal recurrent selection: This is the type of recurrent selection used to improve both general and specific combining ability. This scheme is used to improve polygenic characters. Selection is made on the basis of testcross performance. Two heterozygous populations, each of which is the tester for the other are used in this method.

In the first year, several phenotypically superior plants are selected from the two populations under study. Pollen from the first population is used to pollinate the second population and vice versa. The pollen parents are selfed in both cases. In the second year, evaluation is done for the performance of crosses made in both the populations and the superior crosses are selected. In the third year, the crop from the selfed seeds of the first year is grown. All possible crosses are made within the population. Equal amounts of seed obtained as a result of all intercrosses from the first population and second population are used separately or composited to raise the next generation. This completes one selection cycle which can be repeated.

Recurrent selection is effective in increasing the frequency of desirable genes in the population. It is very much suited for characters with good heritability. The mean values of character shift in the direction of selection. Generally there is no considerable reduction in variability.

5. Family selection: This is a type of recurrent selection in which the unit of selection and recombination is a family. The basic steps of family selection are: (1) development of families (progenies), (2) evaluation and selection of best families, and (3) recombination of families to reconstitute the new base population. Different types of family selection methods have been developed and modified. Mostly half-sib, full-sib or selfed progenies are produced which are evaluated in single or multiple environments. The choice of the type of family selection and evaluation is decided based on the plant species, objectives of selection, heritability of the trait concerned, etc.

C. Disruptive mating and selection

This is a breeding approach in which both the extremes for a character are selected from the segregating population and intermated for the genetic improvement of the population. This can be considered a form of recurrent selection since this method also involves selection and intermating of superior plants.

3.3.3 Selection in Clonally Propagated Crops (Clonal Selection)

The phenotypic characteristics of a plant are the results of the effect of genotype, environment and genotype-environment interaction. Of these, only the effect of genotype is heritable. In the case of clonally propagated plants, the entire genetic structure is inherited as such from the mother plant to the progeny and the variations shown are environmental in nature. A clone is a group of genetically identical plantlets or other propagules developed from a single mother plant without the intervention of sexual recombination. Clonally propagated plants may be non-flowering, low seed setting, apomictic, etc. Nowadays, *in vitro* clonal propagation is being practised in the case of many plants.

Clones are homogenous and genetically identical. In many cases, heterozygous plants with hybrid vigour and such other adaptations are clonally propagated. They show vigorous growth and wider adaptations. Usually they are uniform in behaviour. However, clonal variations are caused by bud mutations, mechanical mixtures and sexual reproduction. Hence, frequent improvement by selection becomes necessary in the improvement of clones.

Clonal selection is the improvement of asexually propagated crops through the method of selection. Superior clones can be isolated from three types of materials: (1) local varieties, (2) introductions, and (3) hybrids.

In the first year, superior plants are selected from a mixed population on the basis of yield, resistance etc. In the second year, the progeny of each selected plant is asexually propagated. Superior clones are selected. In the third year, the selected clones are evaluated in preliminary trials. In the coming 2–3 years, multi-location trials are conducted. Clones superior in yield, resistance and quality are identified. The best clone is multiplied and distributed.

Clonal selection is useful in conserving heterosis. Clones are genetically uniform and therefore stable and uniform in agronomic behaviour. However, new strains of pathogens can easily attack them. Clones will show very little variability and adaptation to new stresses. Clonal degeneration may also occur after a few generations.

3.3.4 Selection in Segregating Populations

Populations like F_2, F_3, etc., are called segregating populations. Techniques of selection like mass selection and pure line selection cannot be applied to such populations. The methods of selection generally used in the case of segregating populations are pedigree method, bulk method and backcross method.

A. Pedigree method

In this method of selection, individual plants are selected in F_2 generation and subsequent generations based on the progeny test. During the entire operation, a pedigree record of the parents and progeny is maintained. Individual plant selection is continued till the progenies show no segregation. At this stage, selection is done among the progenies because there would be no variation within progenies.

> Pedigree record: A detailed record of the relationship between the selected parents and their progenies can be called their pedigree record. Pedigree can be defined as a description of the ancestor of an individual and it generally goes back to its distant ancestor.

Procedure of pedigree method

Pedigree method of selection is an efficient method of handling segregating generations. The steps involved are described below.

1. F_1 generation

The F_1 plants are grown to produce sufficient F_2 seeds. The F_1 plants are planted and assessed for their behaviour with regard to easily observable characters and the selfed plants that occur accidentally are identified and removed.

2. F_2 generation

It is the most important generation that shows maximum genetic variation and provides the opportunity to select individual plants. A relatively large number of plants are raised and evaluated. Selection of plants is made based on their behaviour. The selected plants are individually harvested and a detailed record is kept with respect to genetic identity of the plants and their agronomic characters.

3. F_3 generation

The progeny rows of the selected F_2 plants are grown along with check variety in the F_3. The plants are liberally evaluated and compared visually with check varieties. The promising individual plants are selected within the retained progenies. It must be ensured that progenies with susceptibility to diseases and pests are rejected.

Plants are individually harvested and a record of genetic and agronomic characters is kept.

4. F_4 generation

The progeny of selected F_3 plants is grown to produce F_4 families. The progeny of one F_3 plant constitutes a family. The family differences become obvious in this generation and the number of families can be reduced by visual selection as well as with the help of pedigree records. The rejection of some F_4 families is also needed if they show some prominent undesirable characteristics.

5. F_5 generation

In this generation, the selected plants are grown in progeny rows usually in large numbers. By this generation, variation within families considerably decreases. Progenies or even families showing appreciable plant to plant variations are rejected to set aside the chance of selecting hybrid plants. The selected families are bulk harvested.

6. F_6–F_8 generations

Preliminary yield trials of the selected families are carried out at this stage. Screening for resistance and quality is also done.

7. F_7–F_{10}/F_{12} generations

Multilocation trials are conducted at this stage. Screening for resistance and quality is carried out.

8. Variety release and seed multiplication

When the above steps are accomplished, the selected families are released as new varieties. These new varieties are multiplied and distributed to farmers.

Merits and limitations of pedigree method: Pedigree method provides the maximum opportunity to the breeder to exercise his skill to select plants with the most desirable gene combinations and it helps to evaluate the selections for their performance over a large number of years. At the final stage of bulking, the pedigree record helps to bulk progenies with similar yield potential, tracing back to the F_2 plants. But, it is a laborious and demanding method in terms of resources and time. It involves extensive data recording. The material is to be evaluated only under optimal environmental conditions which makes off-season operations impossible.

Single seed descent method: This is a modification of pedigree method. This envisages complete exclusion of selection among or within lines until sufficient homozygosity is attained by the F_5 or F_6 generation. Here, a large number of F_2 plants are carried forward by maintaining their lineage through single seed from each plant till the F_6 generation. The seeds of the single plants are then grown in progeny rows for their multiplication and trials.

B. Bulk method

The bulk method is a short-term and labour-saving procedure for the management of segregating generations. This technique is adopted to attain sufficient level of homozygosity prior to the initiation of individual plant selection. The seeds are harvested in bulk from F_2 to F_4 or F_5 generation and individual plant selection is practised only afterwards.

Procedure of bulk method

1. F1 generation

The F_1 plants are grown along with the parents and the selfed progeny, if any, is removed. Sufficient F_2 seeds are harvested.

2. F_2 to F_4 generations

Sufficient number of F_2 plants are grown and harvested in bulk. A composite sample of harvested seeds are collected. The process is repeated in the F_3 and F_4 generations. F_5 seeds are harvested.

3. F_5 generation

The bulked F_5 seeds are planted and desirable plants are selected. Each selected plant is harvested separately.

4. F_6 generation

Single plant progeny rows are grown, uniform and promising progenies selected and the selected progenies are bulk harvested separately.

5. F_7–F_9 generations

Preliminary yield trails are conducted to reduce the number of F_7 progenies. Tests are repeated at different locations in the next generation. In F_9 generation, the last progeny line is selected and released as a new variety.

Merits and limitations of bulk method: This method is comparatively inexpensive and easy. Selection starts at a stage when the population attains homozygosity. Single plant selection is done when the plants (F_5 or F_6) are uniform in general. The population is expected to improve for intangible characters like adaptation to existing environment. There is a chance that a loss of genetic variations may take place since selection is delayed. There is a chance that a large proportion of undesirable material is carried forward.

C. Backcross method

The crossing of an F_1 hybrid with either of its parent is referred to as *backcross*. It is used as a method of breeding to improve an otherwise high yielding and popular variety for a specific character.

A superior variety which needs improvement for a specific character serves as the recipient or recurrent parent. The donor or non-recurrent parent acts as a source of gene(s) to be added to the recurrent parent. Any character with good testability can be transferred in this way. Oligogenes can easily be transferred by this backcross method. However, this method can be employed even in the case of polygenic characters.

The genetic basis of inbreeding: Backcross method involves a type of inbreeding since the F_1 is repeatedly backcrossed with the recipient parent. The F_1 contains 50 percent genes from each of the two parents. Repeated backcrossing of the F_1 with the recurrent parent would systematically increase the proportions of gene from the recurrent parent at the cost of the genes from the donor parent.

The procedure of backcross method

1. Transfer of a single dominant gene

The recurrent (rr) and donor (RR) parents are crossed to produce the F_1 (Rr). The recurrent parent is usually used as the female parent. If a cytoplasmic character is transferred, the recurrent parent is used as the male parent.

In the second season, the F_1 plants are backcrossed to the recurrent parent to produce the first backcross generation (BC_1) seeds.

In the third season, the BC_1 plants which consist of Rr and rr in equal proportions are subjected to screening for the character under transfer. The desirable plants are selected and crossed with the recurrent parent to produce the BC_2 seeds.

In the fourth season, the BC_2 plants are grown and they will also segregate in the 1:1 (Rr:rr) ratio. The dominant plants are selected and backcrossed with the recurrent parent to produce the BC_3 seeds.

In the fifth season, BC_3 plants (Rr:rr) are grown and BC_4 seeds are collected as in the case of the previous generation. In the sixth season, BC_4 plants are grown and evaluated. If only four backcrosses are planned, the desirable BC_4 plants are self-pollinated and BC_4F_2 seeds are produced.

In the seventh season, the BC_4F_2 segregates in the 1:2:1 (RR:Rr:rr) ratio. The recessive plants are eliminated and others are self-pollinated to produce BC_4F_3 seeds.

In the eighth generation, BC_4F_3 plants are raised in progeny rows. Progenies produced by the homzygous (RR) BC_4F_2 plants will be uniform for the desired characters, whereas the progenies produced from Rr plants will show segregation. All the uniform progeny rows with the desired character are bulked and released as the improved variety.

2. Transfer of a recessive gene

In this case, the donor parent will be the recessive and usually it is taken as the male parent. In the first season, the dominant × recessive cross is made and F_1 seeds are collected. In the second season, the F_1 is grown and backcrossed with the recurrent (dominant) parent and the BC_1 seeds are collected. In the third season, the BC_1 is grown and selfed to produce BC_1F_2 seeds. In the fourth season, the BC_1F_2 is grown, the dominant plants are eliminated and the other plants are selected and crossed with the recurrent parent to produce the BC_2. In the fifth season, the BC_2 is crossed with the recurrent parent to produce the BC_3 generation. In the sixth season, the BC_3 is grown; selfed and BC_3F_2 seeds are collected. In the seventh season, the BC_3F_2 is grown, dominant plants are eliminated and others are selected. BC_4 seeds collected from the selected plants. In the eighth generation, the BC_4 plants are grown and selfed to produce the BC_4F_2 seeds. In the ninth season, BC_4F_2 is grown and the recessive plants are selected, their seeds bulked and the progeny will be the desired strain of the recurrent parent with the recessive gene from the non-recurrent parent transferred to it.

3. Transfer of a quantitative character

Since more number of loci are involved in the control of a quantitative character and the relative influence of environment in the expression of such traits is higher, a relatively large population is to be sampled in each backcross generation. A general outline of such a procedure is given below.

The F_1 is raised in the first season by crossing the recurrent parent and the donor parent. In the second season, BC_1 is produced. Sufficient quantity of crossed seeds is produced so as to facilitate selection in the BC_1 itself. In the third season, the BC_1 plants are evaluated and those with high intensity of the character under transfer are selected since they carry the maximum number of genes for the character. BC_1F_2 seeds are produced by selfing.

In the fourth season, the BC_1F_2 is raised and is evaluated both for the character under transfer and the general features of the recurrent parent. Plants resembling the recurrent parent with the highest intensity of the character under transfer are selected and harvested individually.

In the fifth season, BC_1F_3 is raised in progeny rows from the above seeds and evaluated extensively for the character under transfer and general features of the recurrent parent. The selected

F_3 plants are crossed with the recurrent parent to produce BC_3 seeds. In the seventh season, the BC_3 plants are grown and selected for the character under transfer and the general feature of the recurrent parent. BC_3F_1, BC_1F_2 and BC_1F_3 generations are then produced in the coming seasons. At the end, a number of lines resembling the recurrent parent and showing the character under transfer are bulked to constitute the modified variety of the recurrent parent.

Merits and limitations of backcross method: This method is very effective to transfer a desired trait to an otherwise improved variety. Evaluation of the variety can be limited to the extent of confirming the transfer of the character under consideration. It provides the ideal solution to utilize the unique properties of an otherwise unadapted germplasm. The limitation of backcross method is that the procedure is time consuming and it does not result in the improvement of other characters.

3.4 PLANT HYBRIDIZATION

Genotypic peculiarities of two or more different varieties or species of plants could be brought together only by crossing them. The mating or crossing of two plants or lines of dissimilar genotype is called *hybridization*. In plant hybridization, one of the plants is taken as the female parent and the other as the male plant. Pollen grains from the male parent are made to pollinate the stigma of the flowers of the female parent. The seeds obtained from such a cross are called *F_1 seeds* and the progeny raised from it is called F_1 (first filial) generation. The F_1 is selfed to produce F_2 and the subsequent generations like F_3, F_4 etc. are raised in the same way. These generations are called *segregating generations* and they are handled differently based on the scope and objectives of the breeding programme.

3.4.1 History of Hybridization

Around 700 BC, the Babylonians and Assyrians used to cross-pollinate date palms as desired. During the recent centuries, hybridization is developed into a scientific method of crop improvement. In 1717, Thomas Fairchild produced the first artificial hybrid, now known as the *Fairchild's mule*, crossing Sweet William variety of *Dianthus* with carnation variety. Joseph Koelreuter (1760-1766) performed several crosses in tobacco and emphasized the occurrence of hybrid vigour in F_1. Gregor Mendel (1865) formulated the principles of inheritance of characters based on his hybridization experiments in *Pisum sativum*. Towards the end of nineteenth century, hybridization was widely used for crop improvement. Now, with the help of the principles of genetics, plant breeders develop new hybrids and varieties for different needs.

3.4.2 Objectives of Hybridization

The major objectives of hybridization are described below:

A. Combination breeding

Combination breeding is the transfer of one or more characters from other varieties to a particular variety. These characters may be oligogenic or polygenic. In this way, genes for disease resistance, quality traits etc. can be transferred.

B. Transgressive breeding

Transgressive breeding is based on transgressive variation of characters in segregating generations like F_2. A cross is made between two strains of plants and the F_2 is screened for transgressive variations.

> Transgressive segregation is the segregation of characters beyond the parental limits, in the segregating generations like F_2.

C. Production of hybrids

Hybrid (F_1) plants show higher vigour and yield when compared to parents, in some cases. This phenomenon is called hybrid vigour. F_1 seeds can be raised in bulk through hybridization and distributed directly for cultivation, especially in cross-pollinating crops.

3.4.3 Types of Hybridization

Based on the genetic difference between parents, hybridization can be classified into: (1) inter-varietal hybridization and (2) distant hybridization.

A. Inter-varietal hybridization

The cross between the members of the same species (intra-specific). is called inter-varietal hybridization. In this type of hybridization, different cross patterns can be used.

1. Simple cross: In this case, two parents are used to produce an F_1 hybrid.

2. Complex crosses: In complex crosses, more than two parents are involved. Such crosses can be called *convergent crosses* since they bring genes from different sources together.

B. Distant hybridization

Hybridization between the members of different species or hybridization beyond species level is called *distant hybridization*. Thus, it may be inter-specific (intra-generic) or inter-generic. When conventional methods of hybridization fails, para-sexual methods are used in such cases.

> Parasexual hybridization is the technique of fusing somatic protoplasts when reproductive cells fail to fuse or fertilize.

3.4.4 The Process of Hybridization

The major steps involved in the process of hybridization are given below:

I. Selection of parents

The choice of parents depends on the objective of the cross. In combination breeding, the genetic diversity of the parents is not important. In the case of transgressive breeding, genetically diverse plants are selected as parents. If the characteristics of the parents are not completely known, they are evaluated for the agronomic features.

II. Emasculation

In the case of crops with bisexual flowers, stamens of the flowers of the female parents are removed or the pollen grains are killed. This process is called *emasculation*. Mechanical, physiological or genetic methods of emasculation are used, depending upon the nature of the crop and the cross.

1. Mechanical methods of emasculation: Here, the anthers are removed from the flowers of the female parents. Hand emasculation and suction method are generally used. For hand emasculation, the flower buds are opened carefully before anthesis (first opening of the flower) and the anthers are removed with the help of forceps. Care should be taken so that the gynoecium of the flowers is not damaged. In suction method, the petals are removed from the flowers before anthesis, with the help of forceps. Then, a thin rubber or glass tube attached to a suction hose is used to suck the anthers from the flowers (Figure 3.2).

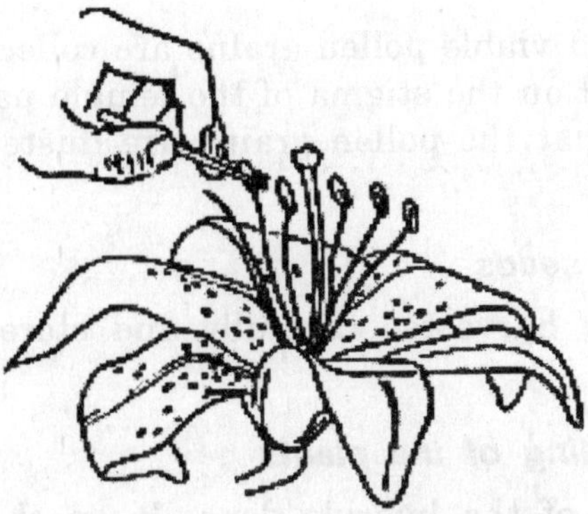

Figure 3.2 Mechanical emasculation.

2. Physiological methods of emasculation: Here, the anthers are killed with the help of heat treatment, cold treatment or alcohol treatment.

3. Genetic emasculation: Genetic or cytoplasmic male sterility factors are introduced into the female parents to make them sterile.

III. Bagging

The emasculated inflorescences of female plants are covered using butter paper bags or cloth bags. However, in the case of cross-pollinated crops, male plants may also be bagged if desired, so as to avoid pollen mixture. The bags are removed 2–3 days after pollination.

IV. Tagging

Emasculated flowers are tagged properly after bagging. Circular or rectangular tags may be used. Details of the cross, date of emasculation, date of pollination and the number of flowers emasculated must be noted in the tag. Carbon pencil or indelible ink may be used for tagging.

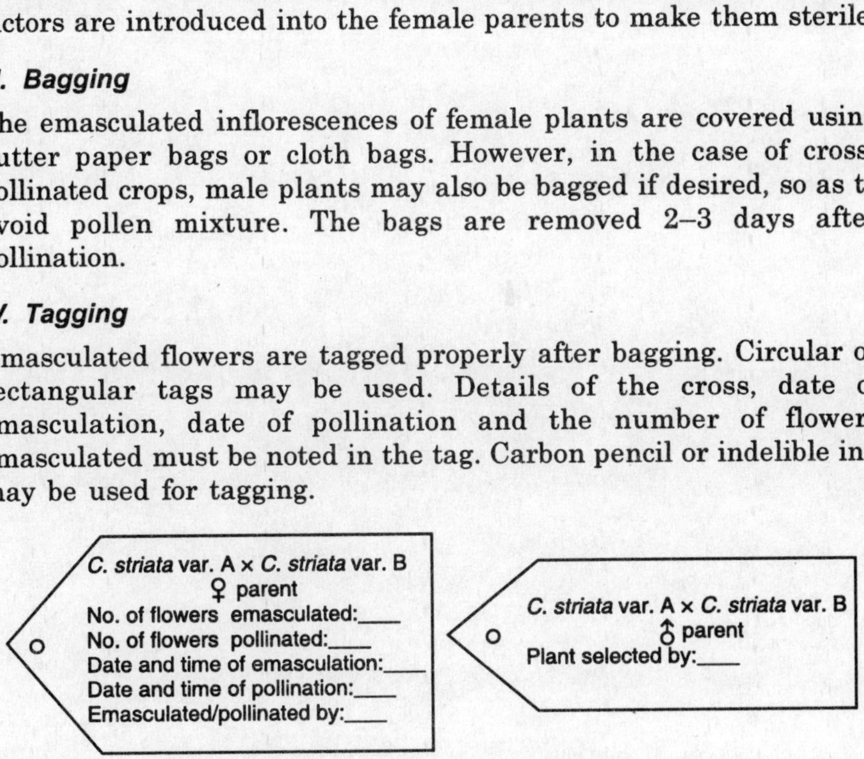

V. Pollination

Mature, fertile and viable pollen grains are collected from the male parent and dusted on the stigma of the female parent. Care should be taken to see that the pollen grains are dusted at the optimum stage of viability.

VI. Harvesting F_1 seeds

Crossed seeds are harvested carefully and stored to raise the F_1 generation.

VII. Further handling of the plants

Further handling of the hybrids depends on the objective of the cross. In the case of hybrid seed production, the F_1 seeds are directly released to farmers. In the case of combination breeding and transgressive breeding, F_2 is raised and the most appropriate solution programme is used.

CHAPTER 4

Modern Methods of Plant Breeding

Conventional methods of plant breeding are age-old techniques that were developed by farming communities and improved by refined plant breeders. On the other hand, modern methods are developed by the plant breeders with the help of modern scientific tools. Modern methods of plant breeding include mutation breeding, polyploidy breeding, distant hybridization and biotechnological methods.

4.1 MUTATION BREEDING

Mutations are heritable changes in the phenotypes of organisms. These changes are the results of chemical changes at the level of genes. Such changes are capable of bringing about new and heritable character variations in crop plants and such variations can be selected and used for the establishment of crop varieties with new characters. Mutations occur in nature in very low frequency. Such mutations are called *spontaneous mutations*. However, the frequency of mutations can be increased with the help of certain chemical or physical agents that are called *mutagens* or *mutagenic agents* and mutations induced in this way are called *induced mutations*. Such agents can be used to induce mutations in crop plants and the desirable variations produced in this way can be selected. This approach of plant breeding in which new variations of crops with desirable characters are developed with the help of induced mutations is called *mutation breeding*.

4.1.1 Mutagens and their Mode of Action

Mutagens are the physical or chemical agents used to enhance the frequency of mutations.

A. Physical mutagens

These are physical agents that are capable of inducing mutations. These are different types of radiations. Radiations can be generally classified into two classes based on their energy levels. Radiations with lower energy levels are capable of causing excitations at the level of nitrogen bases of the genetic material and they are called *non-ionizing radiations*. UV light is a good example of non-ionizing radiations. Radiations with high energy level are capable of causing both excitation and ionization at the level of nitrogen bases. They are called *ionizing radiations*. X-rays, gamma rays, alpha particles, beta particles, fast and thermal neutrons, etc. are the examples of ionizing radiations.

> Nitrogen bases are the basic units of DNA, the genetic material and the nature of hereditary characters is determined by the linear arrangement of the nitrogen bases in the DNA strand and character changes by the re-arrangement, addition or deletion of nitrogen bases. There are two groups of nitrogen bases in DNA: purines and pyrimidines. In total there are two *purines* in DNA: adenine and guanine and two *pyrimidines*: cytosine and thymine.

B. Chemical mutagens

There are chemicals that are capable of enhancing the frequency of mutations. The chemical mutagens are mainly classified into three categories on the basis of their nature of action (Table 4.1).

Alkylating agents: They induce mutations by adding an alkyl group (ethyl or methyl group) to the nitrogen bases. The major alkylating agents are Ethyl Methane Sulphonate (EMS), Methyl Methane Sulphonate (MMS), Ethylene Imines (EI) etc. Since the actions of alkylating agents resemble the actions of radiations, they are known as *radiomimetic chemicals*.

Table 4.1 Major chemical mutagens and their mode of action

Class	*Mutagen*	*Action*
Alkylating agents	Ethyl Methane Sulphonate (EMS)	AT ↔ GC transition
	Methyl Methane Sulphonate (MMS)	Transition
	Ethyl Ethane Sulphonate (EES)	GC ↔ AT transition
	Ethylene Imine (EI)	Transition
Base analogues	5 Bromo Uracil (5 BU)	AT ↔ GC transition
	2 Amino Purine (2 AP)	AT ↔ GC transition
Acridine dyes	Acriflavin	Frame shift
	Proflavin	Frame shift
Others	Nitrous Acid (NA)	AT ↔ GC transition
	Hydroxyl Amine (HA)	GC ↔ AT transition
	Sodium Azide (SA)	Transition

Base analogues: These are chemicals analogous to nitrogen bases. They can get incorporated into DNA at the time of replication and can cause wrong base pairing resulting in mutations. 5-bromo uracil and 2-amino purine are the common base analogues used as mutagens.

Acridine dyes: These are chemicals capable of getting inserted between two bases of DNA, thus leading to addition or deletion of single or a few base pairs at the time of replication. Proflavin and acriflavin are two such examples.

C. Other chemical mutagens

Other chemicals like nitrous acid, hydroxylamine and sodium azide are also efficient mutagens.

4.1.2 Types of Induced Mutations

At the molecular level, induced mutations are changes, additions or deletions of nitrogen bases. Accordingly, they can be classified into transitions, transversions and frame shifts.

Transition mutation is the replacement of one purine with another purine or one pyrimidine with another pyrimidine. Transversion mutation is the substitution of a purine by a pyrimidine or vice versa. Whereas, frame shift mutation is the change in the reading frame of the gene by the addition or deletion of nitrogen bases.

4.1.3 Procedure of Mutation Breeding

1. Selection of the material

The first step in mutation breeding is the decision on the nature of variations to be induced and the selection of the appropriate material for mutagen treatment. Depending upon the method of propagation, seeds or other propagules can be selected for treatment. For *in vitro* mutagenesis, callus or similar *in vitro* material is selected.

2. Choice of the mutagen

Based on the nature of mutation to be induced and the knowledge on the nature of action of the mutagen, the appropriate mutagen is selected. Generally, chemicals are preferred for the seed treatment and radiations are preferred for vegetative propagules, pollen etc.

3. Mutagen treatment

In the case of chemical treatment, presoaking materials in water or solutions of some other chemicals enhance the effect of mutagens. This is called pre-treatment. Later, the materials are transferred into the solutions of the mutagen. A concentration close to LD_{50} of the mutagen is considered optimum. In the case of physical mutagens, the source of the mutagen is kept at a safe distance and the treatment is remotely controlled. Gamma ray treatment is carried out in protected experimental areas known as gamma gardens. The duration of treatment is also decided based on the information available.

> LD_{50} is the doze of the mutagen that causes 50 percent mortality of the treated material (any mutagen is toxic to biological systems and it may cause considerable death and deformities).

4. Handling of the mutated populations in the case of seed propagated species

All the germinated seeds are grown to produce the M1 population. Generally, the mutations will be recessive and most of them can be selected only in later generations. However, dominant mutations and pseudo-dominant mutations can be selected in the M_1 itself. The M_1 plants are selfed and the seeds are harvested separately. The M_2 generation is raised from the seeds collected from the M_1 generation. Oligogenic mutations can be selected at this level. Their seeds are grown separately and desirable mutants isolated after necessary trials. Superior and desirable M_2 plants are selected and M_3 seeds are collected. M_3 progenies are raised from the seeds and they are evaluated for breeding behaviour. The seeds of true breeding progenies are bulked together to conduct yield trials. Preliminary yield trials are conducted in the M_4. Co-ordinated yield trails are carried out from M_5 onwards. By M_8 or M_9, the most promising lines are selected and released.

In the case of polygenic traits, inferior plants are rejected at M_3 and M_4 levels and based on screening tests, the remaining seeds are bulked and used for yield trials and finally released as new varieties.

5. *Handling of mutated populations in the case of clonally propagated species*

In vegetatively propagated species, mutations are expressed as chimeras. Chimeras are combinations of genetically different tissues. Chimeras may be periclinal or sectional. Periclinal chimeras pass through both the inner and outer layers of the growing tips whereas sectional chimeras are limited to some sections of the outer or inner layers of the growing tips of vegetative propagation. In the case of seed propagated crops also, chimeras formed in the inner layers are important. A plant part with completely changed outer and inner layers can be called a solid mutant or homozygous mutant.

In the case of vegetatively propagated crops, the generation raised from the treated propagules is called the VM_1 generation. Plants showing chimeras can be selected and propagated to produce the VM_2 generation. Solid mutants are identified and selected in VM_2. In VM_3, the mutations identified in VM_2 are confirmed. Preliminary yield trials are carried out in VM_4 and co-ordinated trials from VM_5 onwards. By VM_9 the best line is released as a new variety.

4.1.4 Applications of Mutation Breeding

Mutation breeding can be used to develop improved crop varieties, to induce male sterility, for the production of haploids, to create additional genetic variability, to overcome self-incompatibility and to improve the adaptability of crops.

4.2 POLYPLOIDY BREEDING

In somatic cells, chromosomes are present in homologous pairs whereas, in gametes chromosomes are present in single set. Hence, each organism has two types of chromosome numbers, the somatic chromosome number (2n) and the gametic chromosome number (n). However, each genetic set is formed of either a group of different chromosomes or a few groups of such chromosomes. Hence in some cases, the gametic set consists of a few numbers of identical sets. Here, each of such sets represents a basic set of chromosomes and the number of chromosomes in such a set can be called the basic chromosome number (x). Hence n may be equal to x, 2x, 3x etc. when n = x, the organism is diploid, when n = 2x, the organism is a tetraploid and when n = 3x, it is a hexaploid (2n = 2x, 4x and 6x, respectively). Besides this type of variation, absence or additional presence of individual chromosomes can also be seen in organisms. Such variations can be exploited in plant breeding because they bring about desirable character changes in many cases. Variations in chromosome number can be classified as shown Table 4.2.

Table 4.2 Variations in chromosome number

Type	*Characters*
I. Euploidy	Numerical changes in the entire genome
(a) Monopolidy	Only one set of genome (x)
(b) Hapolidy	Only the haploid (gametic) set of genomes (n)
(c) Diploidy	Two sets of genomes (2x)
(d) Polypoidy	More than 2 sets of genomes (3x onwards)
(i) Tripolidy	3x
(ii) Tetraploidy	4x
(iii) Pentaploidy	5x
(iv) Hexaploidy	6x
II. Aneuploidy	Change in the number of one or a few chromosomes
(a) Hypoploidy	Loss of chromosomes from the diploid set
(i) Monosomy	Loss of one chromosome from the diploid set (2n – 1)
(ii) Nullisomy	Loss of one chromosome pair from the set (2n – 2)
(b) Hyperploidy	Additional presence of chromosomes along with the diploid set
(i) Trisomy	Addition of one chromosome to the set (2n + 1)
(ii) Tetrasomy	Addition of one pair of chromosomes (2n + 1)

4.2.1 Haploidy Breeding

Haploids can be used in many ways in plant breeding. They are useful for the development of purelines and inbred lines and for the production of aneuploids. Purelines can be obtained by chromosome doubling of haploids. Such purelines can be used as cultivars or parents in hybridization.

Production of haploids

Haploids originate spontaneously in small numbers. Haploid production can be induced by inter-specific cross, use of alien cytoplasm, anther culture, pollination with foreign pollen, delayed pollination, use of irradiated pollen, temperature shock, chemical treatment etc.

4.2.2 Autopolyploidy Breeding

Autopolyploidy is the condition in which the same genome (x) is present in an organism more than two times. Autotriploid (3x) and autotetraploid (4x) plants are important in plant breeding.

Autotriploids possess three identical sets of chromosomes. Autotriploidy occurs naturally in low frequency. They can be produced by crossing an autotetraploid (4x) with a diploid of the same species (2x). Triploids are usually sterile and non-seed producing. Hence, autotriploidy breeding is very important in fruit crops like banana, apple, grape, watermelon etc.

Autotetraploids (4x) possess four copies of the same genome. They may arise spontaneously or can be induced by doubling the chromosomes of diploid species by colchicine treatment. Autotetraploids are large and vigorous. Examples of autotetraploid crops are rye, grapes, groundnut, potato and coffee.

4.2.3 Allopolyploidy Breeding

Allopolyploids are polyploids in which more than one genome are present. An allotetraploid is otherwise called *amphidiploid* because it contains two genomes twice ($X_1X_1 + X_2X_2$). There are several allopolyploid crop plants that developed in nature spontaneously. Bread wheat (*Triticum aestivum*) (2n = 6x = 42) is an allohexaploid with three genomes: two A genomes from *Triticum monococcum* (2n = 2x = 14), two B genomes from an unknown progenitor (2n = 2x = 14) and two D genomes from *Triticum tauschi* (2n = 2x = 14). Tetraploid American cotton (*Gossypium hirsutum*: 2n = 4x = 48) developed from two diploid species (*Gossipium herbaceum* var. *africanum:* 2n = 2x = 24, an Old World species and *Gossipium raimondii:* 2n = 2x = 24, a New World species).

In tobacco, the cultivated species *Nicotiana tabaccum* (2n = 4x = 48) is an amphidiploid of *Nicotiana sylvestris* (2n = 2x = 24) and *Nicotiana tomentosa* (2n = 2x = 24) and the cultivated species *Nicotiana rustica* is believed to be an amphidiploid of *Nicotiana paniculata* (2n = 2x = 24) and *Nicotiana undulata* (2n = 2x = 24). In *Brassica*, there are diploid and amphidiploid species. *Brassica napus* is an amphidiploid of *Brassica campestris* (n = x = 10) and *Brassica oleracea* (n = x = 9) and *Brassica juncea* is an amphidiploid of *Brassica campestris* and *Brassica nigra* (n = x = 8).

Production of artificial allopolyploids by inter-specific and inter-generic crosses and subsequent chromosome doubling has been carried out with different levels of success. Chromosome doubling is usually effected by treating the diploids with a chemical known as colchicine. Colchicine (C_{22} H_{25} O_6) is an alkaloid obtained from the seeds of the Liliaceous plant *Colchicum autumnale*. Colchicine is applied in concentrations ranging from 0.01% to 0.5%. It is applied to growing tips, meristematic cells, seeds and buds in aqueous solutions or mixed with lanolin. Duration of treatment varies from 24 hours to 96 hours depending upon the plant species.

Colchicine induced polyploidy is known as colchiploidy. It induces polyploidy by inhibiting spindle formation during cell division. Chromosomes do not get segregated at the time of meiosis, resulting in the production of diploid gametes, which on fusion give rise to polyploid plants.

Applications of allopolyploidy breeding

Allopolyploids can be used to produce new crop species, for inter-specific gene transfer and for bridge crosses. Many artificial allopolyploids have been synthesized in different crops. Raphanobrassica is the first example of intergeneric hybridization in plants. This was developed by Karpenchenko in 1927 by crossing radish (*Raphanus sativus*, n = 9) with cabbage (*Brassica oleracea*, n = 9). He developed an amphidiploid by hybridization and chromosome doubling. He could not combine the agronomical characters of the crops. The hybrid had the roots of cabbage and leaves of radish. However, this experiment proved the feasibility of intergeneric hybridization.

Tetraploid species of wheat and cotton have been produced artificially by interspecific hybridization and induction of amphidiploids. Another significant example of intergeneric hybridization followed by polyploidization is the synthesis of the new cereal triticale. Triticale is a man-made cereal produced by crossing wheat with rye. Triticale combines the winter hardiness and high protein content of rye and the bread-making quality of wheat. Hexaploid and octoploid triticales have been developed in this way (Figure 4.1(a) and (b)). Hexaploid triticale is agronomically superior to octoploid triticale.

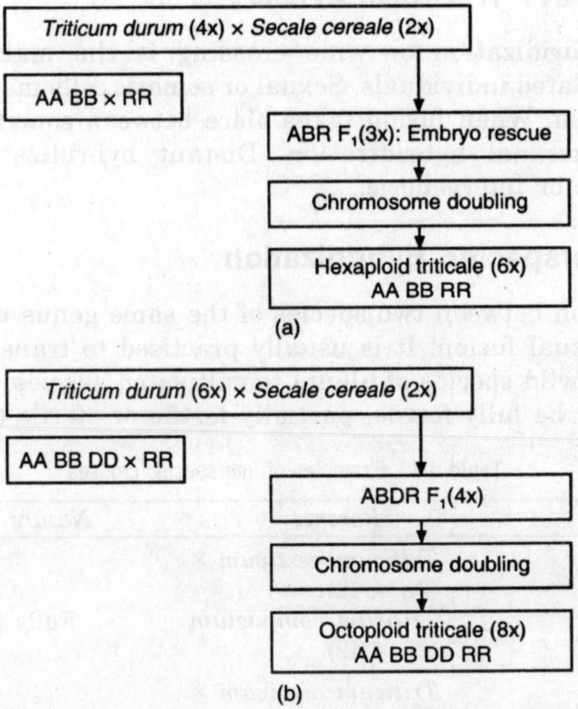

Figure 4.1 Production of (a) hexaploid triticale and (b) octoploid triticale.

> **Bridge crosses:** These are crosses that help in the hybridization between two incompatible species. First, a cross is made between one of the two species and a compatible species. Then the amphidiploid produced in this way is crossed with the second species.

4.2.4 Aneuploidy Breeding

Aneuploids are organisms that show monosomy (2n – 1), nullisomy (2n – 2), trisomy (2n + 1), tetrasomy (2n + 1), etc. They are not directly useful in crop improvement, but they can be used indirectly in different ways. Some of the major uses include locating genes through monosomic and nullisomic analyses; interspecific gene transfer, developing alien addition lines and alien substitution lines of crops and analysis of chromosomal aberrations.

4.3 DISTANT HYBRIDIZATION

Distant hybridization or wide crossing is the mating between distantly related individuals. Sexual or somatic cells may be involved in this fusion. When fusion takes place between somatic cells, it is called parasexual hybridization. Distant hybridization may be interspecific or intergeneric.

4.3.1 Interspecific Hybridization

Hybridization between two species of the same genus usually takes place by sexual fusion. It is usually practised to transfer desirable genes from wild species of plants to cultivated species. Interspecific crosses may be fully fertile, partially fertile or sterile (Table 4.3).

Table 4.3 Examples of interspecific crosses

Crop	*Parents*	*Nature of the* F_1
Wheat	*Triticum aestivum* × (2n = 42) *Triticum compactum* (2n = 42)	Fully fertile
	Triticum aestivum × (2n = 42) *Triticum durum* (2n = 28)	Partially fertile
Cotton	*Gossypium hirsutum* × (2n = 52) *Gossypium barbadense* (2n = 52)	Fully fertile
	Gossypium arboreum × (2n = 26) *Gossypium herbaceum* (2n = 26)	Fully fertile
	Gossypium hirsutum × (2n = 52) *Gossypium thurberi* (2n = 26)	Partially fertile
	Gossypium arboreum × (2n = 26) *Gossypium thurberi* (2n = 26)	Fully fertile
	Nicotiana digluta × (2n = 72)	

(Contd.)

Table 4.3 Examples of interspecific crosses *(Contd.)*

Crop	*Parents*	*Nature of the F_1*
Tobacco	*Nicotiana tabaccum* (2n = 48)	Partially fertile
	Nicotiana sylvestris × (2n = 24) *Nicotiana tomentosa* (2n = 24)	Fully fertile
Brassica	*Brassica nigra* × (2n = 16) *Brassica oleracea* (2n = 18)	Sterile
	Brassica nigra × (2n = 16) *Brassica campestris* (2n = 20)	Sterile
	Brassica oleracea × (2n = 18) *Brassica campestris* (2n = 20)	Sterile

Interspecific crosses help in introgressive hybridization which is the transfer of some genes from one species into the genome of another species. Fertility level of interspecific crosses depends on the homology of chromosomes in the parental species. In the case of sterile crosses amphidiploidy is induced with colchicine and the fertility is restored.

4.3.2 Intergeneric Hybridization

This refers to crosses between two different genera of the same family. Such crosses are not commonly used in crop improvement. However, such crosses may become desirable in a number of situations. Intergeneric crosses can be used when the desirable genes are not present in the same genus, but they are present in allied genera. F_1 hybrids of this type of crosses are always sterile. However, they can be made fertile by chromosome doubling. Intergeneric hybridization has been used successfully in the development of the synthetic cereal, for example, triticale (Table 4.4).

Techniques to make wide crosses successful

1. Selection of parents

The most compatible parents available should be selected for the crosses.

Table 4.4 Examples of intergeneric crosses

Crop produced	*Cross*	*Remarks*
Rabbage (Raphanobrassica) (2n = 36)	Raddish × Cabbage [*Raphanus sativus*] (2n = 18) × [*Brassica oleracea*] (2n = 18)	F_1 sterile. Amphidiploid produced by chromosome doubling are fertile.
Triticale	Wheat × Rye [*Triticum aestivum*] (2n = 42) × [*Secale cereale*] (2n = 14)	F_1 sterile. Amphidiploid fertile (2n = 56)
	Wheat × Rye [*Triticum durum*] (2n = 28) × [*Secale cereale*] (2n = 14)	F_1 sterile. Amphidiploid are fertile (2n = 42)

2. Reciprocal crosses

Reciprocal cross may be attempted when one parental combination fails.

3. Manipulation of ploidy

Diploidization of solitary genomes to make them paired will be helpful to make the cross fertile.

4. Bridge crosses

When two parents are incompatibile, a third parent that is compatible with both the parents can be used for bridge crosses and thus it becomes possible to perform cross between the original parents.

5. Use of pollen mixture

Unfavourable interaction between pollen and pistil in the case of wide crosses can be overcome to some extent by using pollen mixture.

6. Manipulation of pistil

Decapitation of the style will sometimes prove helpful in overcoming incompatibility.

7. Use of growth regulators

Pollen tube growth can be accelerated by using growth hormones like 1AA, NAA, 2,4-D and Gibberellic acid.

8. Protoplant fusion

When fusion of gametes fails, protoplast fusion of somatic cells can be attempted.

9. Embryo rescue

Hybrid zygotes formed by wide crosses may fail to grow in a number of cases. The zygotes are taken out and grown in *in vitro* medium to overcome this problem.

Problems associated with wide crosses

The major problems associated with wide cross are cross incompatibility, hybrid inviability, hybrid sterility and hybrid breakdown.

A. Cross incompatibility: This is the inability of the pollen grains of one species or genus to effect fertilization in another species or genes. This is overcome by employing different techniques like reciprocal crosses, bridge crosses, using pollen mixtures, pistil manipulations, use of growth regulators etc.

B. Hybrid inviability: This refers to the inviability of the hybrid zygote or embryo. In some cases, zygote formation occurs, but further development of the zygote is arrested. In some other cases, after the completion of the initial stages of development, the embryo gets aborted. The reasons for this are: (1) unfavourable interaction between the chromosomes of the two species, (2) interaction between the cytoplasm of the two species, or (3) unfavourable interaction of the maternal tissues and endosperm with the embryo.

Reciprocal crosses, application of growth hormones and embryo rescue are the techniques that can be used to overcome this problem.

C. Hybrid sterility: This refers to the inability of a hybrid to produce viable offspring. This is more prominent in the case of intergeneric crosses. The major reason for hybrid sterility is the lack of structural homology between the chromosomes of the two species. This may lead to meiotic abnormalities like chromosome scattering, chromosome extension, lagging of chromosome in the anaphase, formation of anaphase bridge, development of chromosome rings and chains, and irregular and unequal anaphase separations. These irregularities may lead to aberrations in chromosome structure. Lack of homology between chromosomes may also lead to incomplete pairing of chromosomes. Sterility caused by structural differences between the chromosomes of two species can be overcome by amphidiploidization using colchicine.

D. Hybrid breakdown: This is the condition in which the F_1 is vigorous and fertile but the F_2 is weak and sterile. Hybrid breakdown may be due to the structural difference of chromosomes or problems in gene combinations.

Role of wide crosses in crop improvement

Wide crosses are generally used to improve crop varieties for disease resistance, pest resistance, stress resistance, quality, adaptation, yield etc. These crosses can even be used to develop new crop species (Tables 4.5, 4.6 and 4.7). Techniques like alien addition and alien substitution may also be effective.

Alien addition: Addition of chromosomes of a wild species (foreign species) to the normal compliments of a cultivated species.

Alien substitution: Replacement of one pair of chromosomes of a cultivated species with those of a wild donor species.

Table 4.5 Examples of disease and pest resistance breeding through distant hybridization

Crop	*Characters*	*Donor*	*Recipient*
Wheat	Eye spot resistance	*Aegilops ventricosa*	*Triticum aestivum*
	Rust resistance	*Agropyron*	*Triticum aestivum*
Cotton	Jassid resistance	*Gossypium tomentosum*	*Gossypium hirsutum*
	Boll weevil resistance	*Gossypium armourianum*	*Gossypium hirsutum*
	Rust resistance	*Gossypium raimondii*	*Gossypium hirsutum*
Tobacco	Resistance to TMV	*Nicotiana glutinosa*	*Nicotiana tabaccum*

Table 4.6 Examples of quality breeding through distant hybridization

Crop	*Characters*	*Donor*	*Recipient*
Rice	Seed protein	Wild rice	Cultivated rice
Potato	Starch content	Wild species	Cultivated species
Cotton	Fibre length	*Gossypium thurberi*	*Gossypium hirsutum*
Oats	Oil content	*Avena sterilis*	*Avena sativa*

Table 4.7 Examples of new crop species developed through distant hybridization

Crop	*Parent*
Nicotiana digluta	*Nicotiana tabaccum* × *Nicotiana glutinosa*
Triticale	*Triticum aestivum*/*Triticum durum* × *Secale cereale*

4.4 BIOTECHNOLOGICAL APPROACHES IN PLANT BREEDING

Biotechnology is the age-old technology of using biological tools for the improvement of human life. The twentieth century witnessed tremendous advancements in different areas of biotechnology like microbial technology, genetic engineering and *in vitro* culture technology. *In vitro* culture technology, molecular genetics and genetic engineering have contributed new and unique tools and techniques to plant breeding.

4.4.1 Applications of *in vitro* Culture Technology in Plant Breeding

In vitro culture is the culturing of cells, tissues and organs under aseptic laboratory conditions in culture media. The culture medium generally contains the macronutrients and other supplementary materials like micronutrients necessary for plant growth and other supplementary materials like vitamins, amino acids, carbohydrates and growth regulators.

Plant parts known as explants are cultured in the nutrient medium. Explants may be roots, hypocotyls, cotyledons, leaves, shoot apices, nodal segments, anthers, embryos, etc. The explants are surface sterilized with disinfectants like sodium hypochlorite or mercuric chloride, washed with sterile water and cultured in the nutrient media at 25+10°C.

Usually, depending upon the nature of the explant, the nutrient medium and the hormonal combination, development of callus (an undifferentiated mass of tissue) or direct plantlets from the explant take place within 3–4 weeks.

The callus is subcultured after every 3–4 weeks. The subcultured callus is made to differentiate and produce the shoot system and root system, by altering the composition of the cultural medium. Somatic embryogenesis can also be attempted in callus culture. It is the development of embryo-like structures from cell culture. Such somatic embryos can be encapsulated in a suitable matrix like sodium alginate and synthetic seeds can be produced. Synthetic seeds can be stored for several years and used as natural seeds. The major applications of *in vitro* culture technology in plant breeding include micropropagation, somatic embryogenesis, exploitation of somaclonal variation, meristem culture, anther culture, pollen culture, embryo culture, protoplast culture, cryopreservation of germplasm, secondary metabolite production and *in vitro* mutagenesis from plant cell culture.

Micropropagation

This is the bulk production of clonal plants for rapid propagation. It is an important application of tissue culture technology in plant breeding. It is independent of seasonal and regional constraints. The plants produced in this way are true to type, i.e. they resemble parent plants. Rapid multiplication of planting materials of unique plants with disease resistance and good quality can be carried out by this technique. Uniform behaviour of the clonal crop is highly advantageous in terms of agronomic and harvest practices. But the chances of susceptibility to new strains of pathogens and adverse environmental conditions are always associated with such genetically uniform crop populations.

Somatic embryogenesis

Somatic embryogenesis, encapsulation of the somatic embryos and their storage are very significant steps towards the conservation of genetic resources. These somatic embryos can be used for mass propagation also. Somatic embryogenesis is a process by which embryo-like structures develop from the callus (Figure 4.2). Somatic embryogenesis may be direct or indirect based on the absence or presence of a callus stage.

Figure 4.2 Somatic embryogenesis.

In the process of indirect somatic embryogenesis, embryos arise from callus. The process of somatic embryogenesis takes place in two stages: the induction of proembryonic cell masses or proembryos and the development of proembryos to somatic embryos. Induction of proembryonic masses takes place under high auxin content and development of somatic embryos under low auxin content.

Somatic embryosgeneis is best achieved in suspension culture. Somatic embryos are encapsulated in sodium alginate. Dripping of 2% sodium aliginate from a separating funnel into a 100 millimolar calcium nitrate solution can be used for encapsulation. The encapsulated embryos are placed in calcium for 20 minutes to form complex, rinsed in water and stored in closed containers. The coating material is packed with nutrient hormones and biofertilizers, if desired.

Somaclonal variation

In vitro culturing induces different types of variations like gametoclonal variations, somaclonal variations and protoclonal variations. Gametoclonal variations are variations shown by plants which are regenerated by gametic culture techniques like anther culture or ovule culture. Somaclonal variations are observed among plants regenerated from callus cultures of somatic explants and protoclonal variations can be seen in plants regenerated from protoplast derived callus cultures.

Somaclonal variations may be epigenetic or genetic. Epigenetic variations are temporary and reversible. Whereas genetic variations are stable and irreversible. In long-term cultures, considerable rearrangements in the genome at chromosome or gene level occur and these changes are hereditary in nature.

Somaclonal variations occur spontaneously in low frequencies. They can be screened and beneficial ones are isolated. Somaclonal variations can be induced through the application of pathotoxins, herbicides, salts, metabolic inhibitions, or temperature shocks. The rate of variation expected in tissue culture is approximately 15–10% which is very high in comparison to *in vivo* systems.

The plants regenerated from such callus can be tested for heritability, expressivity and stability and subjected to different levels of field trials. Induction of somaclonal variation is a cheap and quick method of inducing variations.

Meristem culture

Meristems are the growing regions of plants. Apical meristem present towards the stem apex of plants is always free from viral pathogens. *In vitro* culture of shoot apical meristem is widely used to clone pathogen free plants.

Anther and pollen culture

Development of haploid plants is necessary to breed genetically true breeding diploids. The best method to raise haploid plants is anther culture and pollen culture. The haploid plants produced in this way are subjected to chromosome doubling so as to produce genetically uniform plants. Haploids can be produced from female gametes also.

Embryo culture

Embryo culture is the culturing of embryos excised from the ovaries at earlier stages of their development. This technique helps to overcome problems associated with embryo development. Embryos are prevented from development by different factors like incompatibility with the female tissue, absence of endosperm etc. Hybrids produced by wide crosses usually fail to develop inside the ovaries of the mother plants. In such cases, the embryos can be rescued (the technique is called embryo rescue) and grown in culture media so as to produce viable progeny.

Protoplast fusion

Wide crosses fail very often due to the cross-incompatibility of gametes. Fusion of the protoplasts obtained from parent plants followed by protoplast culture can be employed to overcome the problem of cross incompatibility. The technique involves isolation of the protoplasts of the desired cells, fusion of the protoplasts, selection of hybrid cells and culture of the hybrid cells.

The protoplasts are isolated from the mesophyll cells of plants or from the suspension cultures. To obtain protoplasts, cell wall of the mesophyll cells or suspension cells is removed either by mechanical or enzymatic method. The removal of cell wall makes the fusion of protoplast easy. The protoplast fusion takes place in three steps: (i) The plasma membranes of the protoplasts come in close contact, (ii) fuse at small localized regions making protoplast bridges and (iii) then the bridges extend and round off forming homo or heterokaryons. One major difference in the case of somatic fusion is that there is equal contribution of cytoplasm by both the parents. Such hybrids can be called *cybrids* and they are very important in the transfer of cytoplasmic characters.

After fusion, the product will be a mixture of parental type cells, homokaryotic fused cells and heterokaryotic fused cells. The fused cells are selected and cultured to produce somatic hybrids.

Cryopreservation of germplasm

This is the conservation of explants of genetic resources under *in vitro* conditions in liquid nitrogen, usually at 196°C. For the purpose, appropriate explants are selected, treated with certain chemicals (cryoprotectants) to make them resistant to chilling shock, cooled and finally stored in liquid nitrogen. The material is taken out only under critical situations and that too only after several years of storage.

Secondary metabolite production

Many plants, especially medicinal plants, are important due to the presence of different secondary metabolites in them. Production of secondary metabolites in bioreactors in which plant tissues are grown *in vitro* can be employed as an alternative method for the bulk production of secondary metabolites.

In vitro *mutagenesis*

Mutations can be induced by the application of chemical or physical mutagens to *in vitro* culture. This technique is called *in vitro* mutagensis. The mutant cells can be selected and regenerated to give rise to commercially useful mutants of crop plants. Using pathotoxins or fungal culture filtrates, resistant cell lines can be screened and selected. Resistant cell lines have been selected in this way from potato, rice, maize, barley, wheat, sugarcane, oats etc. Amino acid analogues like amino ethyl cysteine and 5-methyl tryptophan are used for *in vitro* selection of cell lines with significant increase in amino acid levels which can be utilized for developing crops with improved nutritional quality. *In vitro* cell selection can be used in the development of agronomically useful mutations. This technique can be used to induce *in vitro* genetic variability which can be subjected to selection.

4.4.2 Molecular Approaches in Crop Improvement

The major applications of molecular biology in crop improvement include gene cloning, use of molecular markers and use of termination technologies.

I. Gene cloning

Gene cloning is the rapid multiplication of genes. Reverse transcriptase technology and *taq* polymerase technology (polymerase chain reaction) are mainly used for the purpose.

Reverse transcriptase technology: Reverse transcriptase is an enzyme produced by retroviruses. This enzyme is capable of catalyzing reverse transcription. It is the process of synthesis of DNA from a messenger RNA template. DNA synthesized in this way is called cDNA (complementary DNA). cDNA consists of only coding DNA. cDNA of organisms can be incorporated with appropriate vectors and stored as cDNA library, by introducing them into bacterial colonies.

The DNA of an organism can be stored in the form of genomic library also. This includes all the expressed as well as non-expressed parts of DNA. Here, the DNA is restricted into fragments with the help of endonucleases and the broken fragments are inserted into cloning vectors, which are then introduced into bacterial colonies and stored.

Polymerase chain reaction: PCR is a technique in which DNA duplication is carried out under *in vitro* conditions, so that multiple copies of the required segment of DNA are obtained. The experiment is carried out in 'eppendorf tubes' in which a polymerase enzyme, the four deoxyribonucleotide phosphates of DNA, and a primer (so that the reaction is initiated) are added. When the reaction is initiated, production of copies of DNA flanked by pair of primers takes place in the PCR machine (thermal cycler).

The major steps involved in PCR are the following:

1. Double stranded DNA (ds DNA) or cDNA is denatured by heating at 95–98°C.
2. Synthetic oligonucleotide primers with sequences complementary to target DNA are added in excess to the reaction mixture along with the nucleotides.
3. When the reaction mixture is cooled to 37°C, primers bind to the complementary sequences present on the single stranded template DNA.
4. The thermostable enzyme 'Taq polymerase' obtained from the thermo-tolerant bacterium *Thermus aquaticus* is added to the reaction mixture. It catalyses the extension of the primers at 72°C from the end of the target region by adding appropriate nucleotides to form a complementary DNA sequence.
5. This cycle is continuously repeated to get multiple copies of the target DNA. Theoretically, PCR yields 2^n copies of target DNA (n = number of cycles). Incubation for 3–4 hours (25–30 cycles) results in million copies of target DNA.

II. Marker assisted selection

Beneficial characters of crop plants can be located and selected with the help of some masker genes. This technique is called *marker assisted selection*. This is more frequent in the case of quantitative characters. Three types of markers are usually used for selection: morphological markers, protein markers and DNA markers.

Morphological markers: Some mutant characters of plants have been found to be linked with other easily observable morphological characters and this property has been utilized in constructing conventional linkage maps. Any character showing association with such morphological markers are supposed to be controlled by genes situated near their loci. The major limitation of morphological markers is their limited number and probable distortion or segregation ratios due to other reasons.

Protein markers: Proteins are products of gene action. The product of a gene can be used as a marker for the presence of a gene. Different alleles of a gene may produce different proteins. Sometimes, different forms of a protein with same catalytic activity but with different molecular weight and electrophoretic properties may be produced by different alleles. Such enzymes are called *isozymes*. The difference in enzyme mobility is caused by point mutations resulting in amino acid substitution. The differences in banding patterns observed on electrophoresis can be used for comparison and selection. Isozyme marker alleles can be associated with other characters and selection is practised. Easily assayable isozymes have been widely used for the characterization of germplasm. However, the availability of useful protein markers is a limitation.

DNA markers: DNA is the genetic material. Each chromosome has about 10^8–10^{10} base pairs but only 10% of the genome is actively engaged in translation. The rest of the genetic material remains unnoticed in terms of character expression. The assessment of variation at the DNA level provides a chance to map the genome. Specific regions on the DNA molecule both in the coding as well as non-coding regions can be identified as markers. The assessment of DNA sequence polymorphism is the most attractive application of molecular biology for human welfare. DNA markers are such polymorphic sequences seen in different individuals.

A considerable part of the DNA shows highly repetitive sequences in the non-coding regions of DNA. The comparison of these regions can also provide valuable clues of genetic and evolutionary relationships. The DNA sequence can be cut with the help of restriction endonucleases, analysed with the help of DNA probes and electrophoretic procedures. Accordingly, different DNA-based markers like RFLP, RAPD, SCAR and SSR have been developed.

Restriction Fragment Length Polymorphism (RFLP): This is the technique of comparing the polymorphism of restriction fragments of DNA. A group of enzymes, known as restriction enzymes, is used in RFLP. Each restriction enzyme is capable of identifying a specific site of DNA, usually 4–8 bp in length, at which it cuts both the strands of DNA. The restriction site of one particular restriction enzyme is present at several regions in the genome of an organism. As a result, using an enzyme, the genome DNA can be cut into a number of fragments known as *restriction fragments*. The length of these fragments depends upon the difference between two adjacent restriction sites. Probing, electrophoresis and autoradioassay of these fragments can give an idea of the variation at the level of restriction fragments. Each genotype has a fixed pattern of distribution of fragments for a given enzyme and probe. RFLP refers to this variation in length of the restriction fragments. More than one enzyme and probe can be used so that the variation in genotypes is assessed perfectly. Unique sequence probes are used more frequently so that only restriction fragments complementary to them are identified and represented in the map. However, the technical complexity of this technique has lead to the development of better techniques that are mostly PCR based.

Randomly Amplified Polymorphic DNA (RAPD): RAPD is a PCR-based technique in which a single short oligoprimer is used to amplify random sequences from a complex DNA template such as a plant genome. The products are then separated by electrophoresis and visualized by UV illumination of ethidium bromide stained gels. Polymorphism of amplified sequences corresponds to genetic differences.

Sequence Characterized Amplified Regions (SCAR): A 'scar' represents a specific region of the genome that is amplified by PCR using a pair of specific oligonucleotide primers. These are identified as distinct single bands in agarose gel electrophoresis.

Simple Sequence Repeats (SSR): Different types of repeated sequences of nitrogen bases have been observed in organisms. Such regions of redundant DNA are called *satellites*. Tandem repeats of about 9–100 bp are called *minisatellites* or *variable number of tandom repeats* (VNTR). Repeats with 1–6 bp are called microsatellites or simple sequence repeats (SSR). DNA sequences containing SSRs can be amplified by PCR and SSR variants can be detected by gel electrophoresis. These can be used as molecular markers.

Other techniques like AP-PCR (Arbitrarily Primed PCR), AS-PCR (Allele Specific PCR), CAPS (Cleaved amplified Polymorphic Sequences), DAF (DNA Amplification Fingerprinting), etc. can also be used for comparing DNA variations.

Molecular genetic maps and their applications in plant breeding: Molecular markers serve as landmarks to identify the phylogenetic and parental relationships of crop varieties. Genetic maps can be constructed based on the nature and location of DNA markers. The extent of recombination between the markers can be analyzed so that map distances can be calculated. A complete genetic map requires the coverage of all regions of all the chromosomes in the genome. This is effectively carried out with the help of computer softwares like MAPMAKER.

Molecular markers are useful in constructing high density genetic maps of crops and also for the location of genes in relation to the markers used. Parallelism in gene order (synteny) and evolutionary relationships can be analyzed. Marker-assisted early generation selection of transgressive segregants can increase the speed of developing new varieties. Selective transfer of desirable genes from wild germplasm can be made easy, thus making intogression of genes easy. Mapping of quantitative trait loci (QTL) becomes easy with the help of molecular markers. Gene pyramiding (assembly of the polygenes responsible for a character) becomes easy with the help of molecular markers. DNA fingerprinting and characterization of crop varieties and germplasm can be used as a very effective tool to analyze the genetic variability of the crop genetic resources available.

III. Terminator technologies

The development of genetically modified crop varieties by several multinational biotech companies simultaneous with the enforcement patenting of life forms and intellectual property rights has lead to the development of new technologies to protect transgenic crops. These are generally called *terminator technologies*.

These technologies prevent the farmer level propagation of genetically modified crops by preventing the germination of seeds. Normal development of the embryo is prevented by arresting the formation of some proteins (enzymes) or by promoting the formation of some undesired proteins vital for seed germination. It is effected by introducing a genetic locus capable of such an action into the genome of such crops. The action of the gene is delayed with the help of a blocking sequence. When the blocking sequence is removed, the lethal gene becomes active thus preventing the germination of seeds.

A terminator system patented by USDA and Delta and Pine Land Company (Scott, Mississippi) under the name 'control of gene expression' consists of two gene systems. The gene system I consists of gene A, promoter PA and a blocking sequence in between the two. A recombinase specific excision sequence (LOX sequence) flanks the blocking sequence on either side. Gene A codes for a protein known as RIP (Ribosome inactivating protein) which causes embryo degeneration. But, gene A remains unexpressed due to the presence of the blocking sequence in between gene A and promoter PA.

The gene system II has two genes—gene B and gene C—with their promoters PB and PC. These genes are involved in the regulation of gene A, which is the terminator gene. Gene B encodes for recombinase which is specific to the LOX sequence of gene A and excises it in order to remove the blocking sequence. Removal of the blocking sequence makes gene A active which produces RIP that destroys the embryo. The expression of gene B is prevented by a repressor protein produced by gene C. As a result, gene A remains inactive in the original seed. Even the progeny seeds produced from it will show normal germination. But, gene B has dual control. Gene B can be depressed by the exogenous application of tetracycline. Tetracycline nullifies the repressive effect of gene C and induces gene B to produce recombinase that is capable of removing the LOX sequence and thus activate gene A, resulting in the production of RIP, the terminator protein. Seeds sold to farmers are treated with tetracycline. It gives a normal crop but the seeds harvested are not visible.

4.4.3 Applications of Genetic Engineering in Plant Breeding (Production of Transgenic Plants)

Genetic engineering is the integration of foreign genes in the genomes of organisms. The technology is otherwise called *recombinant DNA technology*. Transgenic plant production has evolved as a new technique of plant breeding, based on the potentials of genetic engineering. Transgenic breeding involves the incorporation of one or a few desired genes into the genome of a crop plant which is superior in all other aspects. The highest advantage is that the recipient genotype is the least disturbed.

Theoretically, any gene can be transferred by transgenesis. But it is important that the coding sequence of the gene should be transferred along with the promoter genes since the sequences determining plant part, time and rate of expression, etc. are encoded in them. Easily observable markers known as reporter genes should be associated with the transferred gene for effective screening of transgenic plants. Thus, transgenesis is the transfer of a package of DNA sequences that can be called a gene construct which consists of the structural gene, promoter gene and the marker locus. Transgenesis can be adopted both for transgenic plant production and the production of transgenic tissues that can be used in bioreactors for the mass production of desirable components. This technique of producing bulk quantities of metabolites in bioreactors, utilizing transgenic cells is called *molecular pharming*.

I. Gene transfer technology

Genetic engineering or transfer of genes into the genomes of organisms can be performed by two methods—Vector mediated gene transfer and Direct gene transfer.

Vector mediated gene transfer: The most important vectors that are used in plant genetic engineering are *Agrobacterium tumefaciens* and *Agrobacterium rhizogenes*. Some viral vectors like Cauliflower Mosaic Virus and Gemini Virus are also useful in the case of plants.

Agrobacterium tumefaciens (the tumour inducing bacterium) causes the formation of crown galls in plants. The bacterial genophore along with the genes present in a plasmid (Ti plasmid) hosted by it is responsible for the crown gall formation. Genes for the recognition of susceptible cells and binding to these cells are located on the bacterial chromosome, whereas the gene transfer of DNA into the plant genome is encoded in the plasmid DNA. The Ti plasmid DNA has two major regions of importance, the T-DNA and the *vir* region. The T-DNA is capable of producing galls in the host, when transferred into the host cell. Overproduction of certain phytohormones is responsible for the production of gall-like growth. The *vir* region (virulence gene) mediates the transfer of T-DNA.

Modified Ti plasmids with disarmed T-DNA regions (regions not capable of causing crown gall production) are made use for gene transferr. The desired foreign gene is incorporated in the T-DNA region of the plasmid and the bacteria are introduced into the host cells under *in vitro* conditions, either by co-culturing with plant cells or through wounded plant organs.

Agrobacterium rhizogenes induces root formation in the host plants. This is due to the presence of an Ri plasmid (root inducing plasmid).

Agrobacterium mediated gene transfer is suitable in dicotyledons. However, in monocotyledons the response is comparatively poor. Recently, there are reports of *Agrobacterium* mediated gene transfer in monocots also. This is important in terms of crop improvement and agronomy since most of the staple food crops are monocotyledons.

Direct gene transfer: To overcome the problems related to vector mediated gene transfer, methods of direct gene transfer have been developed. Direct methods include physico-chemical uptake of DNA, electroporation, liposome encapsulation, microinjection, macro-injection, biolistics, etc.

1. Physico-chemical uptake of DNA
Recombinant plasmid is mixed with host protoplasts in the presence of polyethyl glylcol, polyvinyl alcohol and calcium phosphate. After 15–20 minutes of incubation, the protoplasts are cultured and screened for homogenesis.

2. Electroporation
This is the creation of pores in the plasma membrame with the help of electrical impulses. Reversible breakdown of the membrane allows the entrance of foreign DNA into the cytoplasm.

3. Liposome encapsulation
Liposomes are small lipid bags enclosing plasmids. DNA can be enclosed in lipid vesicles. When mixed with protoplasts, these vesicles penetrate into protoplasts. Lipase activity of protoplasts dissolves these lipid vesicles and finally the DNA is released and integrated with the host genome.

4. Microinjection
This is the technique of injecting foreign DNA into the protoplasts with the help of injection pipettes.

5. Macroinjection
This involves the injection of relatively large quantity of DNA into developing floral tillers, prior to meiosis.

6. Biolistics
This is the technique of particle acceleration in which DNA is projected into host cells with sufficient force so that it gets integrated with the host genome. DNA coated 0.2–0.7 μm gold or tungsten particles are used. Equipments known as particle gun are used for the purpose. Particle gun accelerates particles to penetrate deep into the target tissue. The process is purely physical and hence no biological barriers are involved. This is applicable to plant species. Transgenic plants by biolistics method have been produced in many crops like rice, barley, soybean, sunflower, wheat, sugarcane, barley, etc.

II. Expression of transgenes

Even after successful introduction of a foreign gene into recipient cells, there is a chance that it does not integrate with the host genome. In such a situation, there will be only transient expression of the gene. The activity of the gene declines over the time and eventually disappears. When the introduced gene gets stably integrated with the recipient genome, it is expressed stably and the phenomenon is called stable expression. The stably integrated transgenes are inherited in Mendelian fashion. But there is a possibility that the expressivity of the gene is lost in the course of time. Such a loss of expression is the result of inactivation of the gene. This phenomenon is called *silencing of trangenes*. The major causes of this are methylation of DNA sequences, co-suppression of the gene or even environmental factors. Gene silencing may be transcriptional in which it is caused due to the methylation of promoter region or post transcriptional in which there is methylation in the coding regions of the gene. Proper screening of the transgenic lines and further hybridization of transgenic plants may help to obtain true breeding transgenic plants with stable expression.

III. Thrust areas of transgenic breeding

The major thrust areas of transgenic breeding include engineering for insect resistance, engineering for disease resistance, engineering for herbicide resistance, engineering for stress resistance, engineering for enhanced food quality and engineering for male sterility.

Engineering for insect resistance: This is one of the pioneer thrust areas of transgenic breeding. Introduction of the *Bt* gene to plants is the most important technique of trangenic breeding. The bacterium *Bacillus thuringiensis* is capable of synthesizing an insecticidal protein (Bt toxin) which is toxic to the larvae of lepidopteran insects. A gene known as *Bt* gene is responsible for the production of Bt toxin. The gene, isolated from *Bacillus thuringiensis* subsp. *kurstaki* has been incorporated to plants like tobacco, tomato, rice, potato, cotton and maize. Use of redesigned synthetic *Bt* gene with increased GC content has also been used in some crops to enhance the expression of the gene.

Some insecticidal proteins of plant origin such as lectins, amylase inhibitors and protease inhibitors have also been identified. Genes responsible for the production of these proteins can be cloned and transferred to plants to induce insect resistance.

Engineering for disease resistance: Engineering resistance to viral diseases, fungal diseases and bacterial diseases has been attempted. In the case of engineering for resistance to viral diseases, genes derived from viruses themselves are used. This is called pathogens derived resistance. This is achieved through the transfer of genes for coat protein mediated resistance, satellite RNA mediated resistance and antisense RNA mediated resistance.

Coat proteins mediated resistance has been achieved in tobacco plants against tobacco mosaic virus (TMV), satellite RNA mediated resistance in cucumber against cucumber mosaic virus (CMV) and antisense RNA mediated resistance in tobacco against TMV.

Resistance to fungal diseases is achieved by transferring genes for antifungal proteins or other antifungal compounds such as phytoalexins.

Introducton of chitinase and glucanase genes into crop plants has resulted in achieving fungal resistance. Chitinase gene has been introduced to rice and tobacco. Co-expression of chitinase and glucanase genes in tomato and tobacco has conferred high level of resistance.

Gene transfer for bacterial disease resistance has not been generally successful. However, the introduction of certain genes like T_4 phage to potato and α-thionin gene to barley have resulted in increased resistance to certain pathogenic bacteria.

Engineering for herbicide resistance: Genes inducing herbicide resistance act by way of producing enzymes detoxifying the herbicides or by way of producing enzymes insensitive to the herbicide. Introduction of bar gene closed from the bacterium *Streptomyces hygroscopicus* makes plants resistant to phosphinothricin (ppt) based herbicides and transfer of *aroA* gene from *Salmonella typhimurium* makes them resistant to glyphosate.

Engineering for stress resistance: Resistance to abiotic stresses like frost, drought and salinity can also be achieved through genetic engineering. For frost protection, an antifreeze gene from fish has been transferred to tomato and tobacco. A gene coding for glycerol-3-phosphate acyl transferase has been transferred to tobacco for cold tolerance.

Engineering for food quality: Presence of amino acids and vitamins of high food value indicates the quality of food. Quality breeding via transgenesis has been carried out in some cases to enhance the quality of food. 'Golden rice' has been produced recently by engineering Beta-carotene gene to the rice genome.

Genes to control ripening and increase shelf life have also been introduced by genetic engineering to some crops. Genes to prevent ripening have been introduced to tomato. Phaseolin gene has been transferred to tobacco and alfalfa to improve protein quality.

Engineering for male sterility: Successful experiments have been carried out for the introduction of nuclear male sterility genes and restorer genes into plants.

CHAPTER 5

Breeding for Special Purposes

Plant breeding aims at improving crop varieties for different purposes. Besides attaining overall general improvement, breeding for special purposes like disease and pest resistance, drought and salinity resistance, quality improvement, production of hybrids as well as synthetic and composite varieties and also for the development of ideal plant type for different crops are some other major threshold areas of plant breeding.

5.1 BREEDING FOR DISEASE AND PEST RESISTANCE

Biotic stress like disease and pest attacks are the major constraints in agriculture. Chemical, mechanical and biological methods of pest and disease control bring about only partial solutions to the problem. Developing disease and pest resistant varieties in different crops is the only feasible solutions to this problem.

5.1.1 Mechanisms of Disease Resistance

Three different mechanisms of disease resistance have been reported.

Resistance to the establishment of the pathogen in the tissue

This is a type of true resistance. Pathogens are not given a chance to establish in the host plant.

Resistance to the growth and development of the pathogen already established in the host

Even though the pathogen infects the host and establishes in it, further growth and development of the pathogen is not permitted. This is also a type of true resistance.

Disease resistance is governed by a number of morphological, physiological and biochemical features of host plants. The mechanism of stomatal opening, osmotic pressure, acidity of cell sap, high crude fibre content, high tannin content, higher concentration of phenols, alkaloids, silica and riboflavin, etc. have been found to be associated with disease resistance.

Disease tolerance

This is the ability of the host to grow and reproduce in spite of the establishment of the pathogen in its tissues.

5.1.2 Mechanism of Pest Resistance

Four mechanisms of pest resistance have been reported from different plants.

Non-preference

This refers to the features of the host plant that make it undesirable or unpalatable to the pests. This phenomenon is also known as *antixenosis*. This is due to the presence of non-preferred colour, hairs, odour, taste or orientation of leaves. For example, in cotton, red plant body, smooth leaves, open canopy, thickness and hardness of boll rind, long pedicel, etc. are characters that cause non-preference for jassids.

Antibiosis

Antibiosis is the adverse effect of the host plant on the growth and reproduction of insect pests that feed on the plants. Antibiosis may even lead to death of the pests. Antibiosis is considered to be the true form of pest resistance. In cotton, presence of high level of gossypol, tannins, silica, etc. are the reasons for antibiosis.

Pest tolerance

This is the ability of a variety to produce fairly good yield even under conditions of mild infection. Tolerance is measured in terms of rejuvenation potential (the ability to repair and replace damaged parts), healthy leaf growth, flowering potential and plant vigour. Hybrid cottons show pest tolerance by virtue of their high fruiting potential.

Pest avoidance/pest escape

The ability of a crop variety to escape from pest attack either by earliness or difference in growing season etc. can be called *pest avoidance*. The crop becomes ready and harvestable before the onset of pest attack.

5.1.3 Basis of Pest and Disease Resistance

The major factors that contribute towards resistance are morphological, physiological or biochemical. However, most of them are gene controlled.

I. Genetics of resistance

The genetic factors contributing towards disease and pest resistance may be oligogenic, polygenic or cytoplasmic.

(a) Oligogenic resistance: This is the condition in which resistance is governed by one or a few major genes. The action of each gene is specific, easily identifiable and transferable. This is also called *vertical resistance* or *specific resistance*. Resistance is uniform but limited to only a specific race of the pathogen. Vertical resistance shows discontinuous variation. This type of resistance is usually limited to only a few generations of the crop. It depends on gene for gene relation of the host and pathogen. Vertical resistance can be exploited in different ways such as (1) development of cultivars with individual major genes, (2) development of multilines, (3) development of cultivars with several major genes (gene pyramiding) and (4) gene deployment.

1. *Developing cultivars with individual major genes*
   In this method, one individual major gene for resistance is transferred to an otherwise superior variety through back-crossing.
2. *Development of multilines*
   Multilines are the mixtures of isolines, related lines or unrelated lines. The seeds of these lines are mixed together. In this way, lines with resistance to different races of a pathogen can be mixed together. A variety developed in this way is resistant to many strains of a pathogen. Even when one line is susceptible to the pathogen, the amount of damage caused is minor.

An isoline is a line of plants with identical genotype

3. *Gene pyramiding*
   This is the incorporation of several genes of resistance into a variety.

4. *Gene deployment*
   This is the use of different genes with same actions in developing resistance races for different geographical areas. This can be done if there are many genes in the same area.

(b) Polygenic resistance: In many cases, resistance may be governed by many genes, each contributing a share towards resistance. This is called *polygenic resistance* or *quantitative resistance*. This is also called *horizontal resistance*. This can also be called general resistance or non-specific resistance since the host plant is resistant to all strains of the pathogen.

This shows continuous variation and hence it is difficult to classify genotypes into resistant and susceptible ones. This type of resistance shows only low heritability. But, it gives protection from different races of a pest. Resistance cannot be easily overcome by new races of pathogens. However, it is difficult to transfer polygenic resistance to new varieties. The mechanism of action is not usually based on gene for gene relationship.

(c) Cytoplasmic resistance: There are examples in which disease (Table 5.1) and pest resistance (Table 5.2) are controlled by cytoplasmic factors. Resistance to southern corn leaf blight in maize, resistance to yellow virus in sugar beet etc. are controlled cytoplasmically.

II. Gene-for-gene relationship

This is a concept developed by Flor in 1956. This concept states that for each gene that centrol resistance in the host there is a corresponding gene that control pathogenicity in the pathogen.

5.1.4 Sources of Resistance

Resistance towards pests and disease may be obtained from different sources like other cultivated species, germplasm collections and mutations and from some microorganisms.

Some varieties of cultivated plants are resistant to many diseases while some others are susceptible. Resistance genes identified in such resistant varieties can be transferred to susceptible varieties so that they become resistant.

Germplasm collections of crop plants including their wild relatives may contain genes for resistance. These genes can be transferred to susceptible varieties.

Disease resistance can be induced through mutagenesis also. Sometimes, natural mutations may also result in the production of disease resistant plants.

Table 5.1 Genetics of disease resistance in some crop plants

Crop	*Disease*	*Genetics*
Rice	Blast	Oligogenic
	Bacterial blight	Polygenic & Oligogenic
Wheat	Yellow rust	Oligogenic
	Stem rust	Oligogenic
	Leaf rust	Oligogenic
Maize	Southern corn disease	Oligogenic
	Corn rust	Oligogenic
	Maize rust	Polygenic
	Leaf rust	Cytoplasmic
Sorghum	Smut	Oligogenic
Barley	Powdery mildew	Oligogenic
	Yellow dwarf	Oligogenic
Potato	Late blight	Oligogenic
	Bacterial blight	Polygenic
Cotton	Fusarium wilt	Oligogenic
	Bacterial blight	Polygenic
Sugar beet	Downy mildew	Oligogenic
	Cercospora leaf spot	Polygenic
	Curly top virus	Polygenic
Tomato	Fusarium wilt	Oligogenic
	Leaf mould	Oligogenic
	Tomato mosaic	Oligogenic
	Bacterial canker	Polygenic
	Bacterial blight	Polygenic
	Curly top virus	Polygenic
Tobacco	Tobacco mosaic	Oligogenic
	Bacterial blight	Polygenic

Table 5.2 Genetics of pest resistance in some crop plants

Crop	*Disease*	*Genetics*
Rice	Plant hopper	Oligogenic
	Stem borer	Polygenic
Wheat	Green bugs	Oligogenic
	Leaf beetle	Polygenic
Maize	Corn borer	Oligogenic & Cytoplasmic
	Leaf aphid	Polygenic
Cotton	Jassids	Oligogenic
Lettuce	Root aphid	Cytoplasmic

Certain microorganisms also carry resistance genes. Bt gene present in the bacterium *Bacillus thuringiensis* is capable of synthesizing a toxic substance known as *Bt toxin* that paralyses lepidopteran insects. The gene (for example, Bt gene) can °be transferred to the crop plants so as to make them resistant to their pests.

5.1.5 Methods of Resistance Breeding

Different plant breeding methods like introduction, selection, hybridization, mutation breeding and transgenesis can be used for breeding disease and pest resistant varieties of crops. Introduction is practised when resistant varieties are available in other countries. Selection of resistant varieties can be practised when such cultivars are available in the germplasm. Genes from such cultivars and wild relatives of crops can be transferred to desirable varieties by hybridization. Disease and pest resistance can also be induced by the method of mutation breeding. Transgenesis method is employed to transfer resistance genes from distant relatives.

5.1.6 Durability of Resistance

The durability of resistance depends upon four major factors:

I. Formation of new races of the pathogen

Formation of new race of the pathogen may make the varieties susceptible to diseases. However, the chances for susceptibility is less in the case of horizontal resistance.

II. Genetics of resistance

Monogenic resistance is less durable than polygenic resistance (controlled by many genes). In general, horizontal resistance is more durable.

III. Morphological and anatomical characters of the host plant

In the case of pest resistance, the morphological and anatomical adaptations achieved by the transfer of genes contribute towards the durability of resistance significantly. Such adaptations cause non-preference and antibiosis.

IV. Biochemical characters of the host plant

Presence of chemicals like silica and toxic induce resistance to pests.

5.1.7 Screening for Disease Resistance

Screening for disease resistance can be carried out both under laboratory and field conditions. Screening under controlled laboratory/glass house conditions is considered to be more reliable. It is done in pots filled with pathogen rich soil. However, screening under field conditions is also important. For the purpose, disease sick plots are either developed by mixing the soil from another plot attacked with the disease, by adding the remains of diseased plants or by adding laboratory cultured inoculum to the soil. Plants are screened for resistance and the seeds are collected and multiplied.

5.1.8 Screening for Pest Resistance

Screening for pest resistance can be done under field conditions as well as glass house/laboratory conditions. In this method also, field testing is necessary because large number of plants can be tested in the fields.

Field screening is done by planting crops in insect-prone areas, inter planting susceptible and resistant varieties, growing plants in heavy infestation seasons and also by hand inoculation with the eggs of pests.

5.2 BREEDING FOR DROUGHT AND SALINITY RESISTANCE

Drought and salinity are the major abiotic stress factors that affect crop production. Hence breeding for drought resistance and salinity resistance are two important strategies of crop breeding.

5.2.1 Breeding for Drought Resistance

Drought is the condition of soil moisture stress. Drought is very common in arid and semi-arid tropics. The ability of plants to grow and reproduce under conditions of soil moisture stress is called *drought resistance.*

Drought causes reduction in growth and yield of crop plants. It damages chloroplasts and reduces photosynthetic efficiency. As a result of drought, proline level in leaves increases, which can be used as an indicator for water stress. Drought resistance is a physiological property which is genetically controlled. It is associated with various morphological and physiological adaptations. Xerophytic plants are more resistant to drought conditions in comparison to mesophytic plants.

I. The mechanism of drought resistance

Different types of drought resistance mechanisms are shown by crop plants.

(a) Drought escape: This is the condition in which the plants complete their life cycle earlier, thus escaping from the peak of the drought season.

(b) Drought avoidance: Drought avoidance is the ability of plants to maintain a favourable internal water balance even under soil moisture stress conditions. This is achieved by the increase in root growth, reduction in the growth of aerial parts, reduction of photosynthesis and reduced transpiratory loss of water.

(c) Drought tolerance: This is the ability of some crop plants to withstand low tissue water content and to grow normally even under such conditions.

(d) Drought resistance: Drought resistance is the ability of crop plants to perform well under moisture stress conditions. This is in general, a combination of drought avoidance and drought tolerance.

II. Adaptations of drought resistant plants

Drought resistant plants show several morphological and physiological adaptations.

(a) Morphological adaptations: The morphological adaptations shown by drought resistant plants include root, leaf, fruit and seed characters.

1. Root features
   Deep, wide growing and branching root system is an adaptation for drought resistance.
2. Leaf features
   Thick cuticle, waxy leaf surface, small and sunken stomata and reduction in stomatal number are the adaptations of leaves for drought resistance. Rapid closing of stomata under water stress conditions also help in drought resistance.
3. Fruit and seed characters
   Presence of awns in the case of wheat and barley has been found to induce drought resistance by increasing the photosynthetic surface of the spike and increasing seed size.

(b) Physiological adaptations: Drought resistant plants show different physiological adaptations. Earliness is an adaptation to avoid drought. Indeterminate growth habit (producing flowers round the year) is also an adaptation to overcome drought stress. Some crop genotypes are capable of maintaining good photosynthetic rate even under moisture stress. Some others overcome drought by reducing their photosynthetic rate. Small cell size, high concentration of cell sap, increased protoplasmic permeability, high viscosity of cytoplasm, etc. are also some physiological adaptations for drought resistance.

III. Sources of drought resistance

Drought resistant genotypes and genes are available from cultivated varieties and germplasm collections. Wild relatives of cultivated plants are rich sources of drought resistance genes (Table 5.3).

Table 5.3 Wild sources of drought resistance in some crop plants

Crop	*Wild species*
Wheat	*Aegilops kotsehyi*
	Aegilops variabilis
	Aegilops speltoides
	Aegilops umbellulata
	Aegilops squarrosa
Sugar cane	*Saccharum spontaneum*

IV. Methods of drought resistance breeding

Breeding methods like introduction, selection, hybridization and mutation breeding can be used for drought resistance breeding. When drought resistance is available in an exotic variety, it can be introduced. When drought resistance genotypes are available among land races, these genotypes can be selected by pure line selection or mass selection. When resistance genes are available in wild species, hybridization can be used to transfer the genes to cultivated varieties. Mutagenesis may be attempted when resistance genes are not available but in case the resistance genes are present in non-related plants transgenic breeding can be used.

V. Drought hardening

Drought hardening is the improvement of drought tolerance of a variety through different treatments. The treatments for drought hardening can be classified into pre-sowing treatments and post-sowing treatments.

(a) Pre-sowing treatment: This is the treatment of seeds before sowing in order to induce drought hardiness. Seeds are soaked in water for 24 hours and they are sun dried. When these seeds are sown in the field, they show improved drought resistance.

(b) Post-sowing treatment: This is the technique of giving mild drought stress to young seedlings. This may improve drought resisting capacity of the crop.

5.2.2 Breeding for Salinity Resistance

Salinity resistance is the ability of plants to grow in saline soils and to show normal growth under such conditions. Higher concentration of salt in the soil usually affects the growth and development of plants by disturbing their physiological process. The mechanism of salt tolerance is genetically controlled.

I. Sources of salinity resistance

Some strains of crops may show better tolerance towards salinity. Crop varieties grown conventionally under saline conditions might be more tolerant to salt stress. Wild relatives of crops may also carry genes for salinity resistance (Table 5.4).

Table 5.4 Wild sources of salinity resistance in crop plants

Crop	*Name of the wild species*
Wheat	*Agropyron seirpea*
	Agropyron junceiformis
Tomato	*Lycopersicon cheesmanii*

II. Breeding approaches for salt tolerance

Salt tolerant lines could be identified among natural land races. But, they may be poor yielding. Improvement of the salt tolerant lines by selection can be practised. Salt tolerance genes from such plants can be transferred to high yielding cultivars. Mutation breeding can also be attempted to induce salt tolerance.

III. Screening of salt tolerant plants

Salt tolerant plants can be screened by sand culture and solution culture. Screening is usually extended over seasons, so as to select the most tolerant lines.

5.3 BREEDING FOR QUALITY

Production of high yielding varieties has been the major thrust area of plant breeding research. However, breeding for improved quality has become a very important outlook of the plant breeder in recent years. Quality is defined as the superiority of a product. In the case of crops and crop products, quality is mainly of three types.

5.3.1 Market Quality

This refers to the fitness of the product for marketing. It is related to uniformity is shape, size, colour and texture in the case of food. Uniform and superior nature of the end product determines market quality.

5.3.2 Industrial Quality

Industrial quality refers to baking quality, malting quality, crushing quality, canning quality, etc. in the case of food crops. Whereas in other crops, the industrial quality parameters depend upon the use of the product.

5.3.3 Nutritional Quality

Nutritional quality refers to the fitness of a plant product for human or animal consumption in terms of the nutrient contents. In other terms nutritional quality depends on the availability of nutrients in the crop products.

5.3.4 Major Nutrients of Crop Plants/Products

The major nutrients of crop products include carbohydrates, proteins, fats, vitamins and minerals.

I. Carbohydrates

Carbohydrates are the most important source of energy for human beings and other animals. It is mainly obtained from cereals and millets. Carbohydrates are stored in plants in the form of starch and sugars. Starch is the storage material in the case of food grains, tuber crops etc. In plants like sugar cane, the storage form of carbohydrate is sucrose. In fruits also the carbohydrate is stored in the form of sucrose. However, in some fruits, the carbohydrate stored is glucose whereas in others it is fructose.

II. Proteins

Proteins are organic compounds made up of amino acids. There are mainly 21 amino acids present in human nutrition. They can be classified into essential amino acids and non-essential amino acids. The essential amino acids are not synthesized in the human body and their requirement is met through dietary intake. However, the non-essential amino acids are synthesized in the human body (Table 5.5). The quality of protein is determined by the presence of essential amino acids.

Cereals contribute about 50% of the global protein requirement and legumes contribute 20%. The remaining 30% is contributed by animal products. Cereals contain about 10–13% of proteins and legumes contain 18–28%. Four types of proteins namely albumins, globulins, prolamins and glutelins are found in cereals. Pulses contain two types of proteins, viz. albumins and globulins. Cereals contain about 70% prolamins and glutelins and only 30% of albumins and globulins. The major storage protein of legumes is globulins (80%).

Cereal proteins are deficient in the amino acids lysine, tryptophan and threonine. Legume proteins are deficient in methionine and tryptophan. Moreover, legumes contain some toxic substances also. Amino acids that are not seen in cereals/pulses are called limiting amino acids. A mixture of cereals and pulses in 3:1 ratio provides nutritionally balanced diet.

Table 5.5 Major amino acids present in human nutrition

Essential amino acids	*Non-essential amino acids*
Methionine	Cystine
Isoleucine	Cysteine
Leucine	Proline
Lysine	Glycine
Threonine	Serine
Tryptophan	Alanine
Valine	Aspartic acid
Phenylalanine	Hydroxyproline
Histidine	Glutamic acid
Arginine	Norleucine
	Tyrosine

(iii) Fats

Fats are energy rich compounds available from plant and animal sources. They serve as reserve food both in plants and animals. Liquid fats are called oils. Oils act as solvent in the case of fat soluble vitamins.

IV. Vitamins

Vitamins are vital compounds required for the proper growth and development of human beings and other animals. However, they are required in minor quantities. Based on their solubility, they can be classified into fat soluble vitamins (A, D, E and K) and water soluble vitamins (B group and C). They are obtained mainly from plant sources.

V. Minerals

Crops and their products are rich source of mineral salts. Many minerals are essential for the proper growth and development of the human body (Calcium, Phosphorous, Sodium, Potassium, Iron etc.).

5.3.5 Genetics of Nutritional Traits

The inheritance of nutritional quality traits are governed by oligogenes, polygenes or cytoplasmic genes.

I. Oligogenic inheritance of quality traits

High lysin content of sorghum and barley, fatty acid composition of safflower, high beta-carotene content of tomato, etc. are controlled by oligogenes.

II. Polygenic inheritance of quality traits

Protein content of cereals and pulses and seed oil content of maize, sunflower, safflower and sesame, etc. are governed by polygenes.

III. Cytoplasmic inheritance of quality traits

Seed oil content of maize and rapeseed, and fatty acid content of maize, soybean and rapeseed show maternal inheritance along with other types of factors.

5.3.6 Sources of Nutritional Quality Factors

The three major sources of nutritional quality are cultivated varieties and germplasm including wild species. Mutations also can contribute nutritional quality factors (Table 5.6).

Table 5.6 Source of quality traits in crop plants

Crop	*Character*	*Source*
Maize	High lysine content	Spontaneous mutation
Sorghum	High lysine content	Spontaneous and induced mutations
Barley	High lysine content	Spontaneous and induced mutations
Oil palm	Highly unsaturated fatty acid content	Wild species (*Elaeis oliefera*)
Tomato	High Beta-carotene content	Wild species (*Lycopersicon hirsutum*)

5.3.7 Breeding Methods for Quality Breeding

Hybridization followed by different methods of selection is the most effective breeding method. Backcross method, pedigree method, recurrent selection, etc. can be employed for the purpose. Mutation breeding (especially with chemical mutagens) has been found to be effective in inducing better quality in crops like sorghum, barley and maize.

5.3.8 Breeding for Low Toxin Content

Toxic substances like tannins, cyanogenic glycosides, saponins, neurotoxins, steroidal chemicals, etc. are seen in many crop plants (Table 5.7). These toxic substances are harmful to human as well as animal health. Development of varieties with low toxin content is a major objective of quality breeding.

Table 5.7 Some toxic substances found in crops

Crop	*Toxic substance*
Sorghum	Tannins and cyanogenic glucocides
Khesari	Neurotoxin (Lathyrogen)
Soybean	Tripsine inhibitor and goiterogens
Cotton seed	Gossypol
Potato	Steroid alkaloids
Cassava	Cyanogenic glycosides
Yam	Alkaloids
Alfalfa	Saponine

5.3.9 Screening Techniques for Quality Traits

Biochemical analysis is the most appropriate technique to identify nutritional and toxic compounds present in crop plants. Seed oil analysis is done with the help of Nuclear Magnetic Resonance (NMR) or Nuclear Infrared Analysis (NIR).

Protein analysis is done based on infrared reflectance measurement with the help of spectroscopic analysis.

5.4 BREEDING HYBRID VARIETIES

F_1 plants produced by crossing two parents with contrasting characters are called *hybrids*. Hybrids usually show increased vigour and fitness over the parents. This property is called *hybrid vigour* or *heterosis*. Due to this property, hybrid populations are directly used for commercial cultivation in the case of many crops. This type of crop production is more common in cross-pollinated species.

Hybrids usually show increased vigour and productivity over their parents. This may be due to the genetic peculiarities of the hybrids. All the individuals of a hybrid population are genetically similar. It is a homogenous population but all the members will be heterozygous for many loci. They show wider adaptability to environmental changes. They are usually more tolerant to biotic and abiotic stresses.

5.4.1 Heterosis

The term *heterosis* was coined by Shull (1914). It is the superiority of F_1 over its parents. Heterosis leads to superiority in adaptation, yield, resistance and quality over the parents. In some cases, negative heterosis may also result. Negative heterosis may also become desirable in some cases. Examples are characters like plant height, duration of flowering, toxin content, etc. Heterosis is confined to the F_1 generation only. It is governed by nuclear genes and sometimes by cytoplasmic factors. Heterosis is positively associated with specific combining ability because it is associated with dominance variance. Dominance variance is associated with heterozygosity. Deleterious recessive alleles are covered by the favourable effects of dominant genes.

The magnitude of heterosis is governed by the mode of pollination, genetic distance between parents, genetic base of the parents and adaptability of the parents.

I. Genetic basis of heterosis

The reason for heterosis is a matter of discussion. Three possible causes of heterosis are: (1) dominance, (2) overdominance, and (3) epistasis.

(a) Dominance hypothesis: This was proposed by Davenport (1908), Bruce (1910) and Keeble and Pellew (1910). According to this hypothesis, heterosis is the result of accumulation of dominant alleles in the hybrid. There is a chance that more number of loci becomes heterozygous and dominant in a hybrid when the alleles from both the parents come together.

For example, if one of the parents is AABBccdd and the other parent aabb CCDD with dominant alleles at two loci each, the hybrid will be AaBbCcDd with dominant alleles at four loci. This may result in increased vigour of F_1. Dominance hypothesis is the most widely accepted one.

Parents: AABBcddd × aabbCCDD
(2 dominant loci) (2 dominant loci)
F_1: AaBbCcDd
(4 dominant loci)

(b) Overdominance hypothesis: This theory was proposed by Shull (1908) and East (1908). According to this hypothesis, heterosis is the result of overdominance of dominant alleles in the heterozygous condition. East (1936) elaborated this hypothesis by proposing a series of alleles of gradually increasing divergence function. Heterosis results due to complementation between divergent alleles.

Parents: AABBCCDD × aabbccdd
(all loci homozygous)
F_1: AaBbCcDd
(all loci heterozygous)

(c) Epistasis: Epistasis is the interaction between alleles of two or more different loci. Nonallelic interaction may be of additive × additive, dominance × dominance or additive × dominance types. Heterosis has got positive association with non-allelic interactions, particularly dominance × dominance interactions.

II. Fixation of heterosis

Heterosis is a property shown by the F_1 and it is difficult to fix it. However, four methods are usually used to fix heterosis.

(a) Asexual reproduction: Vegetative propogation is an effective method to conserve heterosis. This method can be used in sugar cane, potato, sweet potato, banana, etc.

(b) Apomixis: Apomictic progeny developed from maternal diploid cells usually conserve hybrid vigour, since the progeny is identical to the mother plant. Fixation of heterosis by apomixis is possible in citrus, rose etc.

(c) Balanced lethal system: Certain types of balanced lethal systems, in which homozygous organisms do not survive, result in the survival of populations consisting of heterozygous organisms only. The heterozygous organisms of F_1 are usually as vigorous as F_1. This type of fixation of heterosis has been reported from *Oenothera*.

(d) Polyploidy: Heterosis can be fixed by chromosome doubling or polyploidization, especially in interspecific and intergeneric hybrids.

III. Estimation of heterosis

Heterosis can be estimated at three levels.

(a) Average heterosis: This is the change in vigour over the mid parent values.

$$\text{Average heterosis} = \frac{F_1 \text{ value} - \text{Mid parent value} \times 100}{\text{Mid parent value}}$$

(b) Heterobeltiosis: This is the increase in vigour above the better parent value.

$$\text{Heterobeltiosis} = \frac{F_1 \text{ value} - \text{Better parent value} \times 100}{\text{Better parent value}}$$

(c) Useful heterosis: This is the superiority of F_1 over the standard commercial check variety. It is also called *economic* *heterosis*. This is of direct practical value in plant breeding.

$$\text{Useful heterosis} = \frac{F_1 \text{ value} - \text{Commercial check value} \times 100}{\text{Commercial check value}}$$

5.4.2 Breeding of Hybrid Varieties

There are different steps in the breeding of hybrids.

I. Development of inbreds

Development of inbreds is the first step in hybrid production. *Inbreds* are homozygous lines that can be used as parents for hybrid production. Inbreds can be produced either by selfing of heterozygous populations or by doubling of haploids. Superior plants are selected on the basis of vigour, resistance and yield and then selfed. Selfing of superior plants selected from such selfed populations is continued for 6–7 generations to get superior homozygous inbreds. The purpose of selfing is to fix desirable genes in homozygous condition.

II. Selection of productive inbreds

The inbreds are evaluated for their performance on the basis of their general combining ability and specific combining ability. Top-cross method or single-cross method is used to measure the combining ability of inbreds.

III. Production of hybrid seeds

The superior inbred lines are used for the production of hybrid seeds. Single-cross hybrids, three-way-cross hybrids and double-cross hybrids can be produced. Depending upon the crop species, an appropriate technique of emasculation is selected. Cytoplasmic or genetic male sterility can be used to make the female inbreds male sterile. The cost of hybrid seed production can be considerably reduced by this method.

5.4.3 Uses and Achievements of Heterosis Breeding

Heterosis can be exploited fully or partially. Full heterosis is exploited through the development of hybrids and partial heterosis by developing synthetic varieties or composite varieties. In cross-pollinated crops like maize and pearl millet, heterosis is utilized fully as well as partially. But in open cross-pollinated crops and self-pollinated crops, heterosis is fully exploited.

Heterosis has been commercially exploited in many cross-pollinated species like maize, pearl millet, castor, sunflower, cucurbits, onion etc. It has also been exploited in self-pollinated crops like sorghum, tomato, cotton, rice and wheat.

All the present-day cultivated varieties of sugarcane and potato are hybrids. In China, hybrid rice has been developed for commercial cultivation through the use of cytoplasmic-genetic male sterility.

5.5 BREEDING SYNTHETIC VARIETIES

A variety that is developed by inter-mating a number of inbred lines with good general combining ability in all possible combinations and mixing the F_1 seeds in equal quantity is called a synthetic variety. After release, synthetic varieties are maintained by open pollination.

5.5.1 Characteristics of Synthetic Varieties

Synthetic varieties are produced in cross-pollinated crops. These are produced for the partial exploitation of heterosis. A synthetic variety can be developed from inbreds, clones and open pollinated varieties. The end products of recurrent selection which are already tested for GCA are generally used to constitute synthetic varieties. Generally, 5–8 good general combining inbreds are used to constitute a synthetic variety.

The basic concept in developing a synthetic variety is the exploitation of hybrid vigour. However, heterosis is partially utilized here since some quantum of inbreeding takes place due to open pollination in later generations. Synthetic varieties show added vigour mainly due to additive gene action.

A synthetic variety is an assembly of several heterozygotes. But, since it is maintained by open pollination, some degree of selfing occurs resulting in the fixation of some genes. As a result, a synthetic variety gradually becomes a heterogeneous population. The plants are polymorphic and stable and hence highly adapted to environmental variations. These plants can produce stable yield in fluctuating environments. A synthetic variety can be grown for 4–5 generations without reduction in yield potential. The potential can then be maintained by mass selection.

5.5.2 Procedure of Breeding Synthetic Varieties

The major steps involved in the development of synthetic varieties are described below.

I. Isolation of inbreds

Superior inbred lines are developed by recurrent selection and are used for the development of synthetic varieties.

II. Evaluation of inbreds

The inbred lines are evaluated for general combining ability by top-cross method, polycross method or single-cross method. In top-cross, the inbreds are crossed with a common tester and the progenies are evaluated in replicated trials. In polycross method, the selected inbreds are allowed to intermate by open pollinations in isolation. In single-cross method, all possible single crosses are made among selected inbreds. The inbred lines identified by the above methods are selected for the development of a synthetic variety.

III. Intermating of good general combining inbreds

The above inbred lines selected for general combining ability are crossed in all possible combinations. $n(n-1)/2$ is the number of cross combinations for 'n' number of inbreds.

IV. Mixing of F_1 Seeds

The F_1 seeds obtained as above are mixed in equal quantity to constitute the synthetic variety. This generation (mixed F_1) of the synthetic variety is called the *Syn1 generation*. The seeds are usually multiplied by open pollination to produce the Syn_2 and Syn_3 seeds. At this stage the seeds can be released and distributed.

5.6 BREEDING COMPOSITE VARIETIES

A composite variety is a mixture of genotypes from several sources. It is a seed mixture of advanced generation materials of intervarietal or interracial crosses. Seeds of different genotypes with similar characters obtained from such crosses are mixed, subjected to yield trials and finally released to farmers. The advantage of this method is that heterosis from different sources can be exploited and the seeds can be used for 3–4 generations. These seeds have a broad genetic base and show wide adaptation.

5.6.1 Development of Composite Varieties

The different steps involved in the development of composite varieties are described below.

I. Selection of parental material

Open pollinated cultivars or such other heterozygous materials are used as the parental material.

II. Intermating of selected genotypes

The selected genotypes are intermated in all possible combinations. When 'n' number of genotypes are selected as parents, n(n – 1)/2 crosses are made. The seeds are harvested separately.

III. Evaluation of the crossed progeny

F_1, F_2, F_3 and F_4 generations are raised from the cross. The progeny is evaluated at all stages by comparing with the standard checks. Parents that show no depression in the segregating generations are identified.

IV. Mixing of parental seeds of superior cross

Parental seeds of the superior cross are mixed in equal quantity to develop a composite variety. Seed multiplication is carried out by open pollination.

5.7 IDEOTYPE BREEDING

Ideotype is a biological model which is expected to perform in a predictable manner in a defined environment. A crop ideotype is a plant model that is expected to perform better in a particular environment. The term ideotype was proposed by Donald in 1968. The concept of ideotype changes from crop to crop, climate to climate and the concepts and policies of the community concerned.

Ideotype breeding is the development of a plant type with ideal characters. Plant characters are selected from an agronomic point of view.

5.7.1 Features of Crop Ideotypes

A crop ideotype is based on several morphological as well as physiological traits that make the plant more ideal from the economic point of view. However, the characteristics vary from crop to crop and sometimes even within the crop.

Wheat ideotype

An ideotype of wheat should carry the following traits:

1. Dwarf and strong stem;
2. Erect and broader leaves;
3. Long erect ears; and
4. Presence of awns.

Rice ideotype

Rice ideotype should carry the following traits:

1. Dwarf/semi-dwarf culm;
2. High tillering capacity;
3. Short, erect, thick and broad angled leaves; and
4. Long panicles with good grain number.

Maize ideotype

A maize ideotype should have the following traits:

1. Low tiller number;
2. Large cobs; and
3. Angled leaves.

5.7.2 Method of Ideotype Breeding

The method of ideotype breeding consists of four important steps as described below:

I. Development of the conceptual model

The first step in ideotype breeding consists of the development of the concept of the ideal plant type model for the crop based on the environmental conditions. Morphological as well as physiological characters are taken into consideration for the purpose.

II. Selection of the parental materials

Genotypes that can be used to build the ideal model are selected as the next step. Genotypes with broad genetic base and wider adaptability are usually selected. Genotypes for plant stature, proper crop duration, leaf size, leaf angle, resistance, yield, etc. are selected from the germplasm available.

III. Crossing to incorporate the desirable traits

The superior traits of the different genotypes are pooled together by suitable hybridization programmes.

IV. Selection for the ideal plant type

The plants developed as above are screened for the ideal plant type by suitable selection programmes. Preliminary yield trials, multilocation trials and tests for resistance to biotic and abiotic stresses are carried out. Finally, the most ideal genotype is selected, multiplied and distributed for commercial cultivation.

CHAPTER

6

Variety Release and Seed Certification

New verities of crops developed by different plant breeding techniques are released after proper field trials and analysis and are finally made available to the farmers. Each country has developed its own seed production system. Seed legislations ensure control of quality seed production through pre-marketing control and marketing control. Seed production is controlled by National Seed Corporation (NSC), States Seed Corporations (SSC) and State Seed Certification agencies. The important aspects related to variety release and seed multiplication are discussed below.

6.1 VARIETY RELEASE

Plant breeding is mostly a public concern in India but globally it is dominated by the private sector. Release of a new variety developed by plant breeders is done according to certain standard procedures. The varieties are subjected to preliminary yield trials and multiplication trials before their release. The plant breeders present the details of the variety to the authorized committee for evaluation. Usually, the details are presented to the state variety release committee concerned for clearance.

6.2 SEED MULTIPLICATION

Bulk production of seeds is carried out after the achievement of clearance by the variety release committee. Before the production of certified seeds for commercial purpose different stages of seed multiplications are to be carried out.

6.2.1 Breeder's Seed

It is the seed stock or clonal stock directly in the control of the plant breeder. From this stock, other seed types are later developed.

6.2.2 Foundation Seed

It is the first generation of seeds multiplied from the breeder seed. It is done under the supervision of the breeder and seed certification agencies. Foundation seed is the source of all certified seed classers. Foundation seed is produced by Agricultural universities, governmental farms and agencies like NSC and SSC. Foundation seed will be genetically pure because seed multiplication is done with all precautions as in the breeder seed.

6.2.3 Registered Seed

It is the seed produced from foundation seed. This step is performed in the fields of progressive farmers under the supervision of seed production and certification agencies. Registered seed is also genetically pure.

6.2.4 Certified Seed

It is the seed produced from registered seed and released for commercial cultivation. Certified seed is also produced by progressive farmers under the supervision of seed certification agencies.

6.3 SEED CERTIFICATION

Seed certification is the process by which the availability of pure quality seed to farmers is ensured in India. According to Seeds Act, 1966, state seed corporations carry out seed certification. The most important seed standards relate to analytical purity, presence of noxious seeds, germination capacity, moisture content and seed health, minimum seed certification standards have been fixed as applicable to all crops. The two main operations of seed certification are field inspection and seed tests.

6.3.1 Field Inspection

The seed production plot is inspected by the seed certification inspector. To eliminate the chances of mixtures and impurities, field inspections are usually conducted at three stages: before flowering, after flowering and at maturity. The field should be free from mixtures, weeds, pests and durian. The certification agency issues a certificate after each field inspection. Harvested seeds are produced, bagged, tagged and sealed.

6.3.2 Seed Testing

A sample from the seed plot is tested in seed testing laboratories for purity, germination, viability and moisture content.

Purity test

Purity test relates to the comparison of the seeds to the variety under certification. Genetic purity is tested based on progeny performance.

Germination test

Germination test is carried out in seeds. Seeds are germinated on petri dishes.

Viability tests

Viability is tested by tetrazolium test. Tetrazolium is a colourless salt that turns red on reduction by dehydrogenase enzyme. The development of red colour on treatment of seeds with this salt indicates viability. The seed is soaked in water and cut in such a way that enzyme is present in each part. The seed part is soaked in 1% igneous solutions of tetrazolium chloride for four hours. The germinated seeds are counted and the viability percentage is calculated.

Test for moisture content

Seed moisture level is to be maintained at the optimum. For healthy storage and germination, moisture meter is used for moisture testing.

6.3.3 Labelling

Labelling and use of tags of different colour are mandatory in the marketing of commercial seeds. Certified seed is labelled with blue tags, foundation seeds with white tags and breeder seeds with golden yellow tags.

CHAPTER

7

Intellectual Property Rights and Plant Breeding

Agriculture evolved in the hands of farmers for thousands of years. They developed and exchanged seeds and other planting materials suited for their needs and the agroecological conditions of their farms. However, twentieth century witnessed tremendous progress in the field of agricultural production and improvement of agricultural plant varieties. Crop improvement became a modern and sophisticated tool in the hands of specially trained experts. This brought about sustainability and self-reliance in crop production in the case of many developing countries including India. However, the farmer developed varieties were neglected and even when such varieties were used for cultivation, the farmers who developed them were not given any credit. Moreover, the competition between seed producing companies is also acute and, as a result, protection of patent rights and intellectual property rights in relation to crop varieties has become the need of the day.

7.1 PLANT BREEDERS' RIGHTS (PBR)

Competitions in the field of industrial production resulted in the formulation of patent laws. The Indian patenting systems ensured patent protection to industrial process, at the same time providing opportunities for the development of end product by different processes. But the recent amendments shift the stress to product patenting.

Biological systems seeds, vegetation propagate, etc. were free from any such right protection but recent charges of patent laws on a global bars lead to a system of protecting the intellectual property rights (IPR) of the agencies involved in biological researches. Intellectual property rights related to plant breeding are called plant breeders' rights (PBR). The Plant Patent Act of USA, passed on May 12, 1930, was the first legal protection granted to actually propagated plants. In 1970, serenely propagated plants were also made patentable by the plant variety protections Act of USA.

Different forms of Plant Breeders' Rights have been adopted by different European countries. Netherlands adopted the plant Breeders' Decree (PBD) in 1941. An international body, UPOV (Union Pour la protection des ostentations Vegetates—International Convention for the Protection of New Varieties of Plant) was established in 1961 by eight European Countries. UPOV formulated uniform standards for providing plant protection for plant varieties. Presently, UPOV has 43 members. The new version of UPOV standard that was released in 1991 melds all generation and species of plants, propagates and harvested material. Most of the world countries have presently implemented one or other form of PBR. A unified system of PBR is being formalized. The essential criteria for providing patent protection are novelty in distinctions, uniformity and stability.

The issue of PBR in Indian context is a matter of debate. As per the requirements of GATT, India signed an agreement to provide breeders' and farmers' exemption under its IPR system. The Indian Patent Act (1970) has been amended in 1998 and 2005.

The concept of eventually derived variety has been introduced. An amendment towards the curtailment of farmers' exemption was also suggested. The curtailment of farmers' exemption forces farmers to purchase seeds every year from certified source. The definition of essentially desired variety represents a variety developed from a protected variety by adding one or a few genes by backcross method, mutagenesis or transgenesis, but retaining the characteristics of the original variety. Such varieties (essentially derived varieties) qualify for legal protection but these varieties should only be commercialized with the consent of the owner of the original varieties.

GATT has been now replaced by WTO, an international organization to promote liberalization of world trade and to integrate agricultural and trade related property. The agreement on TRIPS

(Trade Related Intellectual Property Rights) entered into force as a part of Uruguay Round of talks on January, 1995, which applies to 134 members of WTO. TRIPS is the first intellectual property law agreement covering agricultural biotechnology to include agriculture and also to permit patenting of products. According to the requirements of WTO, a sui generis system for protection of plant variety has also been included.

7.1.1 Advantages and Drawbacks of PBR

The advantage of PBR is that the breeder gets the patent right of these materials and thus they get benefited. Private investment in plant breeding research will be encouraged. Development of good quality planting materials will be promoted.

On the other hand, agriculture will become an expensive affair because the farmers will loose the right to develop, save and exchange seeds. Patents will restrict free flow of germplasm. Monopoly of private companies in the field of plant breeding and seed production will affect poor farmers adversely. Plant breeding will become profit oriented and the sustainability of agriculture will not be taken into consideration.

7.2 FARMERS' RIGHTS

Plant Breeders' Rights provide legal protection to the right of breeders. However, many of the farmers use farm derived seeds for raising crops. In the process, they even develop new varieties with distinct and novel qualities. Protection of the farmers' rights on the seeds developed in this way has been envisaged in the form of farmers' rights.

7.3 FORMS OF IPR

7.3.1 Plant Patents

The Plant Patent Act, 1930 of USA provides protection to asexually propagated varieties other than those produced by tubers and uncultivated plants. A variety that can be patented under this act should be new and distinct. The requirement of a variety to be considered 'new' is that the variety did not exist previously in a capacity to reproduce itself. Distinctness is made on the basis of easily distinguishable character. There must be human intervention in the process of development of the variety.

7.3.2 Protection under Plant Variety Protection Act (PVPA)

It is the protection given to sexually reproduced varieties of seed bearing plants excluding first generation hybrid varieties. This type of protection is also given from an extension of the Plant Patent Act, 1930 of USA. A variety given protection by PVPA should be novel, distinct, uniform and stable (DUS). The Act prohibits sexual multiplication of a variety and also for using it to produce another variety. But it does not prohibit other breeders from using the protected variety distinctly different from the protected variety. The Act does not prohibit the farmer to save and grow the seeds for the next crop or even to sell the seeds, provided his primary occupation is not seed protection and sale. The Act thus has a provision for breeders' exemption. It provides all the relevant rights to the owner for a period of 18 years.

7.3.3 Utility Patents

Utility patents are industrial patents. It is granted for inventions and products, that are novel, industrially applicable, non-obvious and subject to enabling disclosures. Extension of this type of patenting to plant materials and process opened up scope of utility patents for plant materials. The most notable requirements of utility patenting are enablement disclosure and non-obviousness. The enablement requirement demands a written description of the mention in clear, concise and exact terms. The requirement of non-obviousness demands the plant not to be an obvious variant of known art. These patents give protection to a breed category of plants or plant parts. Unlike PVPA there is no provision for breeders' exemption or farmers' exemption. Utility patents offer substantial breeder protection for plants than is available under PPA or PVPA. Products of nature are not given patents.

7.3.4 Trade Secrets

A trade secret is information that is kept secret to give competitive advantage to the owner over those who do not know it or use it. The nature of the parents in hybrid production can be kept secret in this way. The breeder does not need any patent as the hybrid remains under his control. This type of protection gives unlimited monopoly for invention.

CHAPTER

8

Sustainable Agriculture

Sustainable agriculture is a system of agricultural production that improves the underlying productivity of natural resources and cropping systems so that farmers can meet increasing levels of demands. It is a system that produces food that is safe, wholesome and maintains and promotes human well-being. It ensures adequate net farm income to support an accessible standard of living. It complexes with community norms and meets social expectations. A sustainable farmer practices the following management techniques as a part of the integrated approach towards farming.

8.1 MANAGEMENT TECHNIQUES IN SUSTAINABLE AGRICULTURE

8.1.1 Soil Fertility

In the sustainable approach, the farmer takes a closer look at soil health and evaluate soil biological activities. A healthy soil has a loose crumbly texture, higher level of organic matter and plenty of earthworms. Soil test results combined with farmers' knowledge and plant tissue analysis would provide for the best use of farm resources.

8.1.2 On Farm Resource Utility

Here, the farmer uses his own farm resources before he purchases other inputs. Crop debris and animal manure are used as sources of crop nutrient. Leguminous plants are also integrated into the cropping systems as a natural nitrogen supplement.

8.1.3 Integrated Pest and Disease Management

An integrated approach is maintained towards weed, pest and disease management in a sustainable system. Biological control, use of natural pest and disease control methods, etc. are used and chemical method is adapted only when other systems fail. Crop rotation and tillage practices are also done for weed and pest management.

8.1.4 Crop Rotation

The safest method to increase soil fertility and to break pest life cycles is through crop rotation. Integration of legumes into the crop cycle would supply nitrogen for other crops that need it for their growth. The usual practice in rainfed farming regions of tropical India is cereal-legume crop rotation, that is, one cereal crop rotated with one legume crop. Usually, the cereal crop is grown in the rainy season and the legume is grown in the dry season. Rotation of legumes with rice is very common in South India. This technique provides precious nitrogenous manure. Besides this, the legume crop provides valuable protein to the peasant folk. The famous technique of single straw revolution advocated by Fukuoka in Japan is a rotation between rice and legumes.

8.1.5 High Level Management

For sustainable agriculture better management practices are used and use of chemicals is reduced. On time identification and solution of problems through appropriate practices are essential components of a sustainable approach.

8.1.6 Innovations and Technology Transfer

The farmer develops and uses innovative technology and experiences in his own way. One may develop his own crop mix with added advantages like production of surplus N and help conserve soil. Another farmer may device a management scheme to control live stock diseases.

8.1.7 Diversity

Another property of sustainable agriculture is diversity of crops and practices. As a result, people living on farms and in rural communities can develop and select from numerous risk reducing opportunities that did not exist before. Diversification may increase the efficiency of resource use and bring, within easy reach, certain biological benefits.

8.2 TRANSITION TOWARDS SUSTAINABLE AGRICULTURE

When farmers begin to change their farm practices, they may face a transition period in which a part of the farm undergoes a biological adjustment to the new way of doing things. Fears of lower yields, greater management requirement, possibility of lower profits etc. are perceived by farmers to be the barriers to the adoption of sustainable agriculture. Since each farm differs in location, climate, crop, market, etc., there is no model for recommendation. Each farmer must plan strategies suitable to his situation. Planning, preparation and gradual transition are crucial to success. The concept of sustainable agriculture being evolutionary, the realm of change is a more gradual process.

The farmers should cherish our farmlands as precious national resource centres and they would try to make it improved through their efforts. They have to work to refine and improve their relationship with the land. Under the present trend of increasing cost of fertilizers, pesticides etc., the change to sustainable farming would result in considerable reduction in the cost of production, apart from producing healthy food and reducing pollution of soil, water and atmosphere. Soil health would be continuously upgraded through sustainable methods to achieve better yields with lesser inputs as against the gradual deterioration of soil quality through chemical farming.

8.3 ON FARM STRATEGIES

The first step towards sustainable agriculture is a thorough analysis of the existing farming systems and an assessment of available resources. This would identify the weakness and strength of the total operation. This is also crucial indicating when the changes should be made and in what order. The sustainable approach should be started in one field at a time to allow the farmer to get familiar with the new management techniques. Addition of animal manure or any other organic manure will ease the transition from chemical agriculture, because these manures have got the capacity to stimulate microbial processes in solids that have been damaged by heavy chemical application. Alternative production systems should be evaluated. Integrated pest and disease management system should be practised.

A better flow of information from farmers to consumers and vice versa should be maintained. A partnership of farmers and consumers who share the desire for sustainable agriculture should be promoted. Sustainable educational programmes should be introduced to the curriculum of courses on agriculture and plant breeding. Short-term training programmes should also be conducted for farmers and extension workers.

APPENDIX

I

Some Improved Crop Varieties

Sl. No.	Crop	Cultivated species	Varieties
1	Rice	*Oryza sativa* L. (Asian rice) *Oryza glaberrima* (African rice)	IR8, IR20, IR36, Jaya, Mashuri, Sabari, Revathi, Uma, Navara, Kanaka
2	Wheat	*Triticum aestivum* L.	Sarbathi Sonora, Kalyan Sona, Sonalika, Kanchan
3	Maize	*Zea mays*	Deccan 103, Ganga 11, Hemant, Kisan, Jawahar, Vikram
4	Jowar	*Sorghum vulgare*	Swetha, Swathi, CSH 1, CSH 5
5	Sugar cane	*Saccharum* sp.	Neeraj, Damodar
6	Tea	*Camellia sinensis* (L.) O. Kuntze (China tea); *Camellia assamica* (Masters) Wight (Assam type); *Camellia lasiocalyx* (Cambodia type)	UPASI 1, UPASI 10, TTL 1

Sl. No.	Crop	Cultivated species	Varieties
7	Coffee	*Coffea arabica* L. (Arabica coffee); *Coffea canephora* var. *robusta* (Robusta coffee)	Kaveri, S 274
8	Rubber	*Hevea braziliensis*	RRII 414, RRII 430
9	Cardamom	*Elettaria cardamomum* Maton	ICRI 1, ICRI 2, ICRI 5, ICRI 6
10	Coconut	*Cocos nucifera* L.	CGD, COD, ECT, WCT
11	Arecanut	*Areca catechu* L.	South Kanara Local, Mangala, Sumangala, Sreemangala
12	Oil palm	*Elaeis guineensis* Jacq.	Tenera, Dura, Pisifera
13	Cocoa	*Theobroma cacao* L.	Forastero, Criollo
14	Cashew	*Anacardium occidentale* L.	Madakkathara 1, Madakkathara 2
15	Pepper	*Piper nigrum* L.	Panniyur 1, Panniyur 2
16	Vanilla	*Vanilla planifolia* Andrews	–
17	Ginger	*Zingiber officinale* Rosc.	Varada, Wayanad, Rio-de-Janeiro
18	Turmeric	*Curcuma longa* L.	Suvarna, Suguna, Sudharsana, Prabha
19	Clove	*Syzygium aromaticum* (L.) Merr. & Perry	–
20	Nutmeg	*Myristica fragrans* Hoult	–
21	Cinnamon	*Cinnamomum zeylanicum*	–
22	Indian mustard	*Brassica juncea*	–
23	Black mustard	*Brassica nigra*	–
24	Rape	*Brassica campestris*	–
25	Linseed	*Linum usitatissimum* L.	Neelam, Shubhra, Sweta

Sl. No.	*Crop*	*Cultivated species*	*Varieties*
26	Sesame	*Sesamum indicum* L.	Vinayak, Gauri, Soma, Surya
27	Celery	*Apium graveolens* L.	–
28	Saffron	*Crocus sativus*	–
29	Pigeon pea	*Cajanus cajan* L.	Pusa Ageti, Prabhat, Sweta, Vishaka
30	Tomato	*Lycopersicon esculentum* Mill.	Pusa 120, Arka Vikas, Arka Sourabh, Pusa Ruby
31	Garden pea	*Pisum sativum* L.	Arkel, Bonneville, Kinnauri, Palam Priya
32	Cauliflower	*Brassica oleracea* var. *botrytis*	Pusa Deepali, Pusa Synthetic, Pusa Early Synthetic
33	Broccoli	*Brassica oleracea* var. *italica*	Palam Samridhi
34	Onion	*Allium cepa* L.	Arka Niketan, Arka Kalian, Arka Bindu, Pusa White Flat

APPENDIX II

Model Questions

CHAPTER 1

I. Short Answer Type Questions

1. Expand the following:

(i) FAO	(ii) ICRISAT	(iii) RRII	(iv) IRRI
(v) IPGRI	(vi) ICAR	(vii) CRRI	(viii) CCRI
(ix) IISR	(x) NRCC	(xi) CPCRI	(xii) CRCRI
(xiii) UPASI	(xiv) ICRI	(xv) IARI	(xvi) CARI

2. Where are the above institutes located?
3. Name the headquarters of the following commodity boards:
   (i) Coffee Board (ii) Tea Board
   (iii) Rubber Board (iv) Cardamom Board.

II. Essay Type Questions

1. Give an account of the international institutes of crop science research.
2. Discuss different national agencies involved in crop science research in India. Add a note on the major research institutes under their control.

CHAPTER 2

I. Short Answer Type Questions

A. Write short notes on the following:

1. Mendel's monohybrid cross
2. Mendel's dihybrid cross
3. Law of segregation
4. Law of independent assortment
5. Dominance
6. Co-dominance
7. Incomplete dominance
8. Oligogenes
9. Polygenes
10. Continuous characters
11. Discontinuous characters
12. Inbreeding
13. Hybrid vigour
14. Inbreeding depression
15. Random mating
16. Outbreeding
17. Cell cycle
18. Plasmagenes
19. Independent assortment
20. Joint segregation
21. Maternal inheritance
22. Mutations
23. Mutagens
24. Parts of a flower
25. Microspores
26. Megaspores
27. Megagametophyte
28. Microgametophyte
29. Embryo sac
30. Pollination
31. Fertilization
32. Fruit development
33. Seed development
34. Pseudosexual reproduction
35. Apomixis
36. Apospory
37. Apogamy
38. Adventive embryony
39. Diplospory

40. Parthenogenesis
41. Bulb
42. Corm
43. Tuber
44. Rhizome
45. Tiller
46. Runner
47. Grafting
48. Budding
49. Layering
50. Top working
51. Culture media
52. Callus
53. Organogenesis
54. Pollen culture
55. Embryo culture
56. Endosperm culture
57. Random mating
58. Bisexuality
59. Homogamy
60. Cleistogamy
61. Dichogamy
62. Heterostyly
63. Incompatibility
64. Gametophytic incompatibility
65. Sporophytic incompatibility
66. Male sterility
67. Genetic male sterility
68. Cytoplasmic male sterility
69. Cytoplasmic genetic male sterility
70. Experimental designs
71. Replication
72. Border effect
73. Local control
74. Randomization
75. Completely randomized block design
76. Randomized block design
77. Latin square designs
78. Genetic designs

II. Essay Type Questions

1. Write an essay on the inheritance of characters.
2. Describe and differentiate dominance, co-dominance and incomplete dominance.

3. Write a short essay on discontinuous, continuous and cytoplasmic inheritance of characters in crop plants.
4. Describe independent assortment, linkage and pleiotropy with examples.
5. Write an essay on the systems of reproduction in crop plants.
6. Describe the development of male gametophyte and female gametophyte in angiosperms.
7. Explain sexual, pseudosexual and asexual reproduction in crop plants.
8. Discuss budding, grafting and layering.
9. Describe major steps involved in the micropropagation of crop plants.
10. Write an essay on mating systems in crop plants.
11. Write an essay on different systems of incompatibility and sterility seen in crop plants.
12. What are experimental designs? Discuss their significance.

CHAPTER 3

I. Short Answer Type Questions

A. Write short notes on the following:

1. Domestication of plants.
2. Changes of characters under domestication.
3. Natural and artificial selection.
4. Speciation under domestication.
5. Centres of origin for cultivated plants.
6. Name five crops originated in the Hindustan centre.
7. Name five crops originated in the Central American centre.
8. Name two important crops originated in the Abyssinian centre.
9. Name the centres of origin of sugar cane and black pepper.
10. Plant introduction.
11. Primary introduction.
12. Secondary introduction.
13. Agencies of plant introduction.
14. Procedure of plant introduction.
15. Germplasm.
16. Genetic resources.
17. Conservation of genetic resources.
18. *In situ* conservation.

19. *Ex situ* conservation.
20. Field gene banks.
21. *In vitro* conservation.
22. Base collections.
23. Active collections.
24. Working collections.
25. Core collections.
26. Mass selection.
27. Pure line selection.
28. Progeny selection.
29. Simple recurrent selection.
30. Recurrent selection for general combining ability.
31. Recurrent selection for specific combining ability.
32. Reciprocal recurrent selection.
33. Clonal selection.
34. Pedigree method of selection.
35. Bulk method of selection.
36. Backcross method of selection.
37. Plant hybridization.
38. Hybrid.
39. Fairchild's mule.
40. Combination breeding.
41. Transgressive breeding.
42. Types of hybridization.
43. Distant hybridization.
44. Parasexual hybridization.
45. Handling of F_1 population.

II. Essay Type Questions

1. Write an essay on domestication of plants and speciation under domestication.
2. Write an essay on the centres of origin of cultivated plants.
3. Write an essay on plant introduction.
4. Discuss the importance of crop genetic resources and write the major methods of conservation.
5. Write an essay on the methods of selection that can be employed in self-pollinating crops.
6. Write an essay on the methods of selection that can be employed in cross-pollinating crops.
7. Discuss the methods of selection practised in segregating populations.
8. Write an essay on plant hybridization and add notes on the major objectives and techniques of hybridization.

CHAPTER 4

I. Short Answer Type Questions

A. Write short notes on the following:

1. Physical mutagens.
2. Chemical mutagens.
3. Polyploidy breeding.
4. Importance of haploidy breeding.
5. Autopolyploidy in crop improvement.
6. Allopolyploidy in crop improvement.
7. Triticale.
8. Aneuploidy in plant breeding.
9. Distant hybridization.
10. Intergeneric crosses.
11. Problems associated with wide crosses.
12. *In vitro* culturing.
13. Micropropagation.
14. Somatic embryogenesis.
15. Somaclonal variation.
16. Meristem culture.
17. Anther culture.
18. Embryo culture.
19. Protoplast fusion.
20. Secondary metabolite production.
21. *In vitro* mutagenesis.
22. Reverse transcriptase technology.
23. Polymerase chain reaction.
24. Marker assisted selection.
25. Protein markers.
26. Molecular markers.
27. RFLP.
28. RAPD.
29. SCAR.
30. SSR.
31. Terminator technology.
32. Molecular pharming.
33. Direct gene transfer.
34. Vector mediated gene transfer.
35. Biolistics.
36. Expression of transgenes.
37. Genetic engineering for insect resistance.
38. Genetic engineering for disease resistance.

39. Genetic engineering for herbicide resistance.
40. Genetic engineering for stress resistance.
41. Genetic engineering for food quality.

II. Essay Type Questions

1. Write an essay on mutation breeding and its potentials and limitations.
2. Discuss the importance of polyploidy in the evolution and breeding of crops.
3. What is distant hybridization? Discuss its methods and major achievements.
4. What are the major applications of biotechnology in plant breeding?
5. What are molecular markers? How are they useful in crop improvement programmes?
6. Discuss the major applications of recombinant DNA technology in Plant Breeding.

CHAPTER 5

I. Short Answer Type Questions

A. Write short notes on the following:

1. Mechanism of disease resistance.
2. Mechanism of disease and pest resistance.
3. Genetics of disease and pest resistance.
4. Major sources of disease and pest resistance genes.
5. Major methods of resistance breeding.
6. Screening for disease resistance.
7. Screening for pest resistance.
8. Mechanism of drought resistance.
9. Sources of drought resistance genes.
10. Drought hardening.
11. Salinity resistance.
12. Breeding for quality.
13. Food quality of crop plants.
14. Inheritance of quality traits.
15. Breeding for low toxin content.
16. Breeding for hybrid varieties.
17. Heterosis.
18. Genetic basis of heterosis.
19. Fixation of heterosis.

20. Estimation of heterosis.
21. Breeding of hybrids.
22. Synthetic varieties.
23. Composite varieties.
24. Ideotype breeding.

II. Essay Type Questions

1. Explain the major mechanisms of pest and disease resistance in crop plants.
2. Write an essay on the genetics of pest and disease resistance.
3. What are the major methods adopted in breeding for pest and disease resistance?
4. What are the major adaptations for drought and salinity reistance? Give examples.
5. What is quality breeding? Write an essay on the major quality aspects of food crops and their potential for improvement.
6. Explain heterosis and heterosis breeding. How can heterosis be fixed? Write the methods of estimating heterosis.
7. Write an essay on breeding hybrid varieties, synthetic varieties and composite varieties.

CHAPTER 6

I. Short Answer Type Questions

A. Write short notes on the following:

1. Breeder's seed.
2. Foundation seed.
3. Registered seed.
4. Certified seed.

II. Essay Type Question

1. Write an essay on seed certification.
2. Bring out the importance of seed certification in modern agriculture.

CHAPTER 7

I. Short Answer Type Questions

A. Write short notes on the following:

1. Plant breeder's right.
2. Plant patents.
3. Plant Variety Protection Act.
4. Utility patents.

II. Essay Type Question

1. Write an essay on patent rights and their application in agriculture.
2. Discuss the merits and demerits of patenting in agriculture.

CHAPTER 8

I. Short Answer Type Questions

A. Write short notes on the following:

1. Integrated pest and disease management.
2. Crop rotation.
3. Transition towards sustainable agriculture.
4. On farm strategies in sustainable agriculture.

II. Essay Type Question

1. Write an essay on the importance of sustainable agriculture. Add a note on farm strategies in sustainable agriculture.
2. How can you make agriculture sustainable? Write a note on the importance of sustainable agriculture.

Index